The Institute of Mathematics
and its Applications
Conference Series

# The Institute of Mathematics and its Applications Conference Series

Previous volumes in this series were published by Academic Press to whom all enquiries should be addressed. Forthcoming volumes will be published by Oxford University Press throughout the world.

NEW SERIES
1. *Supercomputers and parallel computation* Edited by D. J. Paddon
2. *The mathematical basis of finite element methods*
   Edited by David F. Griffiths
3. *Multigrid methods for integral and differential equations*
   Edited by D. J. Paddon and H. Holstein
4. *Turbulence and diffusion in stable environments* Edited by J. C. R. Hunt
5. *Wave propagation and scattering* Edited by B. J. Uscinski
6. *The mathematics of surfaces* Edited by J. A. Gregory
7. *Numerical methods for fluid dynamics II*
   Edited by K. W. Morton and M. J. Baines
8. *Analysing conflict and its resolution* Edited by P. G. Bennett
9. *The state of the art in numerical analysis*
   Edited by A. Iserles and M. J. D. Powell
10. *Algorithms for approximation* Edited by J. C. Mason and M. G. Cox
11. *The mathematics of surfaces II* Edited by R. R. Martin
12. *Mathematics in signal processing*
    Edited by T. S. Durrani, J. B. Abbiss, J. E. Hudson, R. N. Madan, J. G. McWhirter, and T. A. Moore
13. *Simulation and optimization of large systems*
    Edited by Andrzej J. Osiadacz
14. *Computers in mathematical research*
    Edited by N. M. Stephens and M. P. Thorne
15. *Stably stratified flow and dense gas dispersion*
    Edited by J. S. Puttock
16. *Mathematical modelling in non-destructive testing*
    Edited by Michael Blakemore and George A. Georgiou
17. *Numerical methods for fluid dynamics III*
    Edited by K. W. Morton and M. J. Baines
18. *Mathematics in oil production*
    Edited by Sir Sam Edwards and P. R. King
19. *Mathematics in major accident risk assessment*
    Edited by R. A. Cox
20. *Cryptography and coding*
    Edited by Henry J. Beker and F. C. Piper
21. *Mathematics in remote sensing*
    Edited by S. R. Brooks
22. *Applications of matrix theory*
    Edited by M. J. C. Gover and S. Barnett
23. *The mathematics of surfaces III*
    Edited by D. C. Handscomb
24. *The interface of mathematics and particle physics*
    Edited by D. G. Quillen, G. B. Segal, and Tsou S. T.

# The Interface of Mathematics and Particle Physics

Based on the proceedings of a conference
organized by the Institute of Mathematics and its
Applications on the Interface of Mathematics and
Particle Physics held at the University of Oxford
in September 1988

Edited by

D. G. QUILLEN, G. B. SEGAL, and TSOU S. T.

CLARENDON PRESS · OXFORD · 1990

Oxford University Press, Walton Street, Oxford OX2 6DP
Oxford New York Toronto
Delhi Bombay Calcutta Madras Karachi
Petaling Jaya Singapore Hong Kong Tokyo
Nairobi Dar es Salaam Cape Town
Melbourne Auckland
and associated companies in
Berlin Ibadan

Oxford is a trademark of Oxford University Press

Published in the United States
by Oxford University Press, New York

© The Institute of Mathematics and its Applications, 1990

All rights reserved. No part of this publication may be reproduced,
stored in a retrieval system, or transmitted, in any form or by any means,
electronic, mechanical, photocopying, recording, or otherwise, without
the prior permission of Oxford University Press

This book is sold subject to the condition that it shall not, by way
of trade or otherwise, be lent, re-sold, hired out or otherwise circulated
without the publisher's prior consent in any form of binding or cover
other than that in which it is published and without a similar condition
including this condition being imposed on the subsequent purchaser.

British Library Cataloguing in Publication Data
The Interface of mathematics and particle physics.
1. Theoretical physics
I. Quillen, D. G.   II. Segal, G. B.   III. Tsou, S. T.   IV. Series
530.1
ISBN 0-19-853626-7

Library of Congress Cataloging in Publication Data
(applied for)

Printed in Great Britain by
St. Edmundsbury Press, Bury St. Edmunds, Suffolk

# Preface

These are the proceedings of the IMA Conference on 'Mathematics-Particle Physics Interface', held last September at the Mathematical Institute, Oxford. The proceedings could not have appeared, nor indeed the meeting have taken place, without the help of many whom it is our pleasure to thank here.

We have had, throughout the preparations, the actual meeting, and finally the editing, the advice and support of Professor Sir Michael Atiyah. Rob Baston lent us his organizational skills and especially his expertise in LaTeX. To Jacek Brodzki fell not only the task of taking photographs and so on, but also the tedious work involved in putting manuscripts in LaTeX in the OUP format. He did this with meticulous care and artistry. Lukas Nellen assisted us in sorting out some computing technicalities. Our administrator, Sheila Robinson, was helpful in providing us with the usual services during the meeting despite the summer staff shortage. We are most grateful to them all. The clerical and financial side of the conference was managed by the IMA.

Finally we thank speakers, contributors, chairpersons and participants alike for making possible this conference. We hope that the material that follows will be useful to mathematicians and physicists who are interested in knowing what is happening in the other field.

D.G. Quillen
G.B. Segal
Tsou S.T.

MATHEMATICAL INSTITUTE, OXFORD.

# Contents

1. Quantum Groups and Conformal Field Theories    1
   *L. Alvarez-Gaumé*

2. Essay on Physics and Non-commutative Geometry    9
   *A. Connes*

3. Twistors, Particles, Strings and Links    49
   *R. Penrose*

4. Instantons in Yang–Mills theory    59
   *S.K. Donaldson*

5. Extended Conformal Algebras    77
   *Peter Goddard*

6. Super Riemann Surfaces    87
   *Alice Rogers*

7. Gauge Theories and Relativistic Membranes    97
   *P.K. Townsend*

8. The Space of 2d Quantum Field Theories    107
   *Emil J. Martinec*

9. Chern–Simons Forms and Cyclic Cohomology    117
   *D. Quillen*

10 A Yang-Mills Structure for String Field Theory          135

   *Tsou Sheung Tsun*

11 An Approach to Constructing Rational Conformal Field Theories          143

   *K.S. Narain*

12 A Universal Link Invariant          151

   *R.J. Lawrence*

13 Strings and Quantum Gravity          157

   *H.J. de Vega and N. Sánchez*

14 Quantum Group generalizations of String Theory          161

   *H.J. de Vega and N. Sánchez*

15 Towards a Covariant Closed String Theory          167

   *J.G. Taylor and A. Restuccia*

16 Contact Interactions for Light-cone Superstrings          171

   *A. Restuccia and J.G. Taylor*

17 Closed String Field Theory: the Failure of BRST Cohomology for the Open String          175

   *P. Mansfield*

18 Twistors and Four-dimensional Conformal Field Theory          181

   *M.A. Singer*

19 Projective and Superconformal Structures on Surfaces          187

   *W.J. Harvey*

20 Unified Spin Gauge Theories of the Four Fundamental Forces 193

   *J.S.R.Chisholm and R.S.Farwell*

## Contents

21  Path Integral Formulation of Chiral Gauge Theories     203
    *T.D. Kieu*

22  The Correct Significance of the Binary Pulsar Observations     209
    *D. F. Roscoe*

23  Approaches to Scale Invariance in Two Dimensional Statistical Mechanics     211
    *P.P. Martin*

24  String Amplitudes and Twistor Diagrams: an Analogy     217
    *A. P. Hodges*

25  Lie Cochains on an Algebra     223
    *Jacek Brodzki*

List of Participants     229

# 1

# Quantum Groups and Conformal Field Theories

L. Alvarez-Gaumé

Important advances have been made recently towards the classification of Rational Conformal Field Theories (RCFT). A RCFT is characterized by a chiral algebra $\mathcal{A} = \mathcal{A}_L \times \mathcal{A}_R$ such that $\mathcal{A}_L$ (resp. $\mathcal{A}_R$) contains at least the identity operator and the Virasoro algebra, and the Hilbert space $H$ of the theory splits into a finite number of irreducible representations of $\mathcal{A}$: $H = \bigoplus H_i \times H_{\bar{\imath}}$, with $i, \bar{\imath}$ running over a finite range of values. Examples are provided by the minimal models [1] and the discrete unitary series of Virasoro representations [2] whose chiral algebra is the Virasoro algebra; the two-dimensional Wess–Zumino–Witten theory [3] with $\mathcal{A}_L$ an affine Kac–Moody algebra, etc. A classification of RCFT is important in the determination of universality classes of two-dimensional critical systems and it may also be an important step towards the resolution of the far more difficult problem of understanding the space of classical solutions to string theories.

In [4], Verlinde studied the fusion algebra of RCFT, which is a consequence of the operator algebra of the theory. The structure constants of this algebra is given by the different couplings between three conformal families. If $[\phi_i]$ denotes the conformal family of the primary field $\phi_i$, the fusion algebra is written as:

$$[\phi_i] \times [\phi_j] = \sum_k N_{ij}{}^k [\phi_k], \qquad (1)$$

and the $N_{ij}{}^k$ are non–negative integers. If we define the matrices $(N_i)_j{}^k = N_{ij}{}^k$, the associativity of the operator algebra of the conformal theory implies that the $N_i$'s commute. More abstractly, the fusion algebra is a commutative associative algebra with as many generators as conformal families in the theory and with structure constants $N_{ij}{}^k$. For each family $[\phi_i]$ we can construct its character:

$$\chi_i(\tau) = \text{Tr}_{[\phi_i]} q^{L_0 - c/24}, \qquad q = e^{2\pi i \tau}. \qquad (2)$$

The behaviour of (2) under modular transformations in a modular covariant theory is:

$$
\begin{aligned}
T: \quad & \chi_i(\tau+1) = e^{2\pi i(h_i - c/24)}\chi_i(\tau), \\
S: \quad & \chi_i(-1/\tau) = S_i{}^j \chi_j(\tau),
\end{aligned} \tag{3}
$$

where $h_i$ is the conformal dimension of $\phi_i$, and $c$ is the central extension of the Virasoro algebra. In [4] it was shown in many examples that the matrix $S$ diagonalizes the fusion rules. More precisely, if we write $N_{ij}{}^0 = C_{ij}$ and use $C$ to lower indices, then

$$N_{ijk} = \sum_m \frac{S_{im} S_{jm} S_{km}}{S_{0m}}, \tag{4}$$

and the eigenvalues of $N_i$ are $\lambda_i^{(k)} = S_i{}^k / S_0{}^k$. This striking connection between modular transformation and the fusion rules was proved rigorously in [5] using a set of polynomial equations characterizing RCFT. The polynomial equations involve the matrices $S$ and $T$ and two other matrices $C, N$ expressing the duality properties of the tree-level conformal blocks. To exhibit the equations satisfied by $C$ and $N$ it is convenient to introduce chiral vertices [6]. They are operators which represent the holomorphic three-point functions:

$$\Phi\left\{{}^{\;i}_{jk}\right\}(z) : H_k \to H_j, \tag{5}$$

where $i$ is an index for a primary field. The conformal blocks can be written in terms of expectation values of products of chiral vertices. For example

$$\mathcal{F}_p^{ijkl}(z,w) = \langle i \mid \Phi\left\{{}^{\;j}_{ip}\right\}(z) \Phi\left\{{}^{\;k}_{pl}\right\}(w) \mid l \rangle. \tag{6}$$

The matrix $C$ describes the exchange of two chiral vertices. At the level of vertices it is the matrix representing the braiding (through analytic continuation) of the $j, k$ legs of (6). Graphically:

$$\begin{array}{c} {}^j \quad {}^k \\ \underline{\;\;\;\;}_{i\;\;\;\;\;p\;\;\;\;\;l} \end{array} = \sum_{p'} C_{pp'}\begin{pmatrix} j & k \\ i & l \end{pmatrix} \;\underline{\;\;\;|\;\;\;|\;\;\;} $$

$$\tag{7}$$

# 1. Quantum Groups and Conformal Field Theories

where the left-hand side is a pictorial representation of the block $\mathcal{F}_p^{ijkl}$. The matrix $N$ is a consequence of the associativity of the operator product expansion:

$$\begin{array}{c} j \quad k \\ i \quad p \quad l \end{array} = \sum_{p'} N_{pp'}\begin{pmatrix} j & k \\ i & l \end{pmatrix} \quad \begin{array}{c} j \quad k \\ i \quad p' \quad l \end{array} \tag{8}$$

The hexagon equation for $C$ follows from the defining relations of the braid group, and the pentagon equation satisfied by $N$ is a consequence of the associativity of the operator product expansion (see [5] for details). The proof of the Verlinde conjecture is a consequence of the pentagon identity after one writes a precise representation of the Verlinde operators. These are defined as follows: on the torus we choose a homology basis $(a, b)$. For a RCFT given a primary field $\phi_i$, we can always find a conjugate field $\phi_{\bar{i}}$ such that the operator product $\phi_i \times \phi_{\bar{i}}$ contains the identity. The Verlinde operators $\phi_i(a)$, $\phi_i(b)$ act on the characters $\chi_i(\tau)$. They correspond to inserting the identity factorized into $\phi_i$, $\phi_{\bar{i}}$, then taking $\phi_i$ around the $a$ or $b$ cycle, and finally recombining the two operators into the identity once again. If the $a$-cycle represents the equal-time surface, the action of $\phi_i(a)$ on $\chi_j$ is diagonal:

$$\phi_i(a)\chi_j(\tau) = \lambda_i^{(j)}\chi_j(\tau). \tag{9}$$

The action of $\phi_i(b)$ is more complicated:

$$\phi_i(b)\chi_j(\tau) = A_{ij}{}^k \chi_k(\tau), \tag{10}$$

since $a$ and $b$ are exchanged by the modular transformation $S$, then $\phi_i(b) = S\phi_i(a)S^{-1}$, and the matrices $(A_i)_j^k \equiv A_{ij}{}^k$ all commute. In fact, $S$ diagonalizes the $A_i$'s. The conjecture by Verlinde was that $A_i$ and $N_i$ coincide.

There are two ways that quantum groups [7] enter into conformal field theory. First, the Verlinde operators are associated to closed paths on the torus. It is possible to construct analogues to these operators for open paths at tree level [8]. The advantage is that these operators are compatible with the operation of taking traces, and they also imply the hexagon and pentagon equations, hence the proof the Verlinde conjecture becomes conceptually simpler. Furthermore these open Verlinde operators satisfy the defining relations for a quantum group. The second application of quantum groups has to do with the fact that they provide solutions to the polynomial equations [9]. Moreover, all the known RCFT can be obtained using the GKO construction [10] for appropriate groups $G$, $H$, $H \subset G$, and their $C$, $N$ matrices and modular properties seem to follow from the quantum deformations of $G$ and $H$ [11]. Whether the complete set of solutions to the polynomial equations are given by quantum groups is not clear at present.

In the first application we begin with the space of chiral vertices $V$ compatible with the fusion rules $N_{ij}{}^k$. On $V$ we can define two operations: one is sewing $*$, and the other one is taking characters. We want to find an algebra of automorphisms of $V$ compatible with 1) sewing, 2) braiding, 3) the $N$ operation (8) or s–t duality. Denoting colectively by $s$ the three labels in the chiral vertex (5), braiding (or exchange) operation can be written as

$$\Phi_{s_1}(z)\Phi_{s_2}(w) = R_{s_1 s_2 s_3 s_4}\Phi_{s_3}(w)\Phi_{s_4}(z). \tag{11}$$

If $Q$ is the automorphism of the algebra and $X \in Q$, we can write the action on $V$ as:

$$X(\Phi_{s_1}) = \sum_{s_2} C^{(1)}_{s_1 s_2}\Phi_{s_2}. \tag{12}$$

If we think of $R$ as acting on $V \otimes V$, condition 1) becomes:

$$\begin{aligned} R\, C_1\, C_2 &= C_2\, C_1\, R, \\ C_1 &= C \otimes 1, \quad C_2 = 1 \otimes C, \end{aligned} \tag{13}$$

reminiscent of the definition of a quantum group [7]. If all we had was (13), we could always define a representation of $Q$ via the adjoint representation (in analogy with ordinary Lie algebras). The problem is that the constraint imposed by the fusion rules: $\Phi\left\{{j \atop ik}\right\}\Phi\left\{{m \atop ln}\right\}$ vanishes unless $k = l$. Nevertheless, for every primary field we can construct an element of $Q$. Choosing for simplicity a self-conjugate primary field (i.e. $\phi_a \times \phi_a = 1 + \cdots$) we obtain:

$$X^a(\Phi\left\{{j \atop ik}\right\}) = \sum_{m,n} R\left({a \atop mi}\right)\left({j \atop ik}\right), \left({j \atop mn}\right)\left({a \atop nk}\right)\Phi\left\{{j \atop mn}\right\}. \tag{14}$$

Graphically, we can represent the chiral vertex as in Figure 1, with the representation $H_i$, $H_k$ attached in the open boundaries and the primary field $\phi_j$ at the point $z$. The operator (14) can be interpreted as in Figure 2: we factorize the identity into $\phi_a$, $\phi_{\bar{a}}$ and then move the

# 1. Quantum Groups and Conformal Field Theories

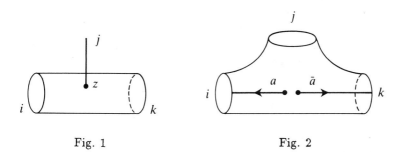

Fig. 1                    Fig. 2

fields to the $i, k$ ends along the path shown. For more complicated surfaces one can do the same operation for each open path. This is why we can interpret $X^a$ as the open Verlinde operators. The conditions 1), 2) and 3) imply also the tree level polynomial equations (see [8] for details). Finally, the action on the characters is based on the operation of taking traces. In particular, the characters are, schematically:

$$\chi_i = \text{tr}_{H_i} q^{L_0} \quad i \quad \begin{array}{c} id \\ | \\ \end{array} \quad i \; , \qquad (15)$$

where we have represented the chiral vertex by $i \; \begin{array}{c} j \\ | \end{array} \; k$. The action of $X^a$ on $\chi_i$ is defined by first acting on the chiral vertex and then taking the trace. It is easy to see, using (14) and the triviality of braiding with the identity, that

$$X^a(\chi_j) = \sum_m N_{aj}{}^k \chi_k, \qquad (16)$$

which is indeed the standard Verlinde operator.

The second use of quantum groups is related to the solutions of the polynomial equations. This can be illustrated with the quantum group $SL(2, q)$ which is associated to the level $k$ Wess–Zumino–Witten theory (WZW) (see [9]). This algebra satisfies the defining relations:

$$[X^+, X^-] = \frac{q^{H/2} - q^{-H/2}}{q^{1/2} - q^{-1/2}}, \quad [H, X^\pm] = \pm 2X^\pm. \qquad (17)$$

When $q$ is an arbitrary real or complex number the representation theory of this algebra is analogous to the classical case. In RCFT, $q$ is a root of unity. For instance, in the WZW theory, $q = \exp(2\pi i/k+2)$. Now the representation theory becomes more interesting. The only regular representations have spin $j = 0, 1/2, 1, \ldots, k/2$, as in the WZW theory. Furthermore, the composition of angular momentum generates the WZW fusion rules:

$$[j_1] \times [j_2] = \sum_{j=|j_1-j_2|}^{\min(j_1+j_2, k-j_1-j_2)} [j], \tag{18}$$

and the $C$ and $N$ matrices are given by $q$-analogues of the 6–j symbols computed in [12]:

$$C_{jj'} \begin{bmatrix} j_2 & j_3 \\ j_1 & j_4 \end{bmatrix} = (-1)^{j+j'-j_1-j_4} q^{(c_{j_1}+c_{j_4}-c_j-c_{j'})/2} \left\{ \begin{matrix} j_2 & j_1 & j \\ j_3 & j_4 & j' \end{matrix} \right\}_q$$

$$N_{jj'} \begin{bmatrix} j_2 & j_3 \\ j_1 & j_4 \end{bmatrix} = \left\{ \begin{matrix} j_1 & j_2 & j \\ j_3 & j_4 & j' \end{matrix} \right\}_q.$$

In this way, one reproduces the results of [6], and using the $q$-characters of $SL(2,q)$, one finds that the modular transformations are represented by $q \mapsto q^{-1}$.

A plausible reason why the quantum group appears is because the polynomial equations are concerned with the Hilbert space of the theory modulo the chiral algebra. In the Kac–Moody case relevant for the WZW theory this means roughly that we forget the moding in the Kac–Moody generators. Naively one might expect after doing this one is left with the classical algebra. The deformation, however, is a consequence of the central extension of the Kac–Moody algebra which is crucial in determining its representations.

It is quite plausible that all solutions to the polynomial equations are given by combinations of quantum groups. Work in this direction is in progress.

### Acknowledgements

I would like to thank the organizers of the meeting for the opportunity to present this material in such a stimulating environment.

### References

[1] A.A. Belavin, A.M. Polyakov and A. Zamolodchikov, *Nucl. Phys.* **B241** (1988) 333.

[2] D.Friedan, Z. Qiu and S. Shenker, *Phys. Rev. Lett.* **52**(1984) 1575.

[3] E. Witten, *Comm. Math. Phys.* **92** (1984)455.

[4] E. Verlinde, *Nucl. Phys.* **B300** (1988)360.

[5] G. Moore and N. Seiberg, *Commun. Math. Phys.* **123** (1989) 177.

[6] See:

B. Schroer, Algebraic Aspects of Non–Perturbative Quantum Field Theories, *Como Lectures*, August 1987, and references therein;

J. Fröhlich, Statistics of Fields, the Yang–Baxter Equations and the Theory of Knots and Links, *Cargèse Lectures*, 1987;

A. Tsuchiya and Y. Kanie, *Lett. Math. Phys.* **13** (1987)303

[7] V. G. Drinfel'd, Quantum Groups in *Proc. of International Congress of Mathematicians,* MSRI, Berkeley, 1986.

[8] L. Alvarez-Gaumé, C. Gomez and G. Sierra, *Nucl. Phys.* **B319** (1989) 155.

[9] L. Alvarez-Gaumé, C. Gomez and G. Sierra, *Phys. Lett.* **B220** (1989) 142.

[10] P. Goddard, A. Kent and D. Olive, *Comm. Math. Phys.* **103** (1986)105.

[11] L. Alvarez-Gaumé, C. Gomez and G. Sierra, in preparation.

[12] A.N. Kirillov and N.Yu. Reshetikhin, *LOMI Preprint* E-9-88.

THEORY DIVISION, CERN, GENEVA, SWITZERLAND.

# 2

# Essay on Physics and Non-commutative Geometry

*A. Connes*

Our aim, in this article, is to try to discover what physics would be like if the space in which it took place was not a set of points, but a non-commutative space. We shall not go very far in this direction, and the consequences of this investigation are for the moment either mathematical (see Sections 3 and 4 below) or only applied to a commutative space-time (Section 5). It is clear, however, that a tool as remarkable as the Dixmier trace for analyzing logarithmic divergences should be useful to physicists. Moreover we have been able to show that a small modification of our picture of space-time (see Fig. 2) gives a conceptual explanation of the Higgs fields and of the way they appear in the Weinberg–Salam model. This should allow us to make at the classical level explicit predictions of the Higgs mass: a very crude one is discussed in model II below.

The general formalism of non-commutative geometry, in which the notion of points gives way to that of coordinates, of functions on the space and of the algebra of function, is analogous to the general formalism of field theory, where the dynamics of point particles subject to an action at a distance gives way to that of fields with a local action. It is therefore natural, given a non-commutative space $X$ (given by an algebra $\mathcal{A}$ of functions on $X$) to start by working out the analogue of a field theory on $X$. To do this we have to: a) identify the classical fields on $X$, b) to write down the analogue of the Lagrangian of the standard model.

In field theory the points $x$ of space play the role of a dummy index only, except by their appearance in the following two rules which experience has shown us to be all important.

1. The principle of gauge invariance of the second kind, which says that the theory admits an invariance group $\mathcal{U}$ of infinite dimension, the group of maps of $X$ into a compact Lie group. It is this invariance which generates the conservation of local currents $j(x)$.

2. The principle of local interactions, which says that, as soon as we go beyond the free field theory, in order to write down physical interaction we have to localize all the free fields present and to write down interaction in a local form, using combinations like $\phi(x)\psi(x)\phi(x)$ of field values at the same point.

It is as a result of the gauge invariance principle that contact with non-commutative geometry is most easily established. In fact, if the compact Lie group $G$ (gauge group of the first kind) is $U(n)$, $n \geq 2$, the gauge group of the second kind $\mathcal{U}$ determines and is determined by the algebra $\mathcal{A}$ of functions on $X$. Moreover, this group is the unitary group of the algebra $M_n(\mathcal{A})$ : $\mathcal{U} = \{u \in M_n(\mathcal{A}); uu^* = u^*u = 1\}$, where $M_n(\mathcal{A})$ is the algebra of matrices on $\mathcal{A}$. Even if the algebra $\mathcal{A}$ was commutative, the algebra $M_n(\mathcal{A})$ is no longer so ($n > 1$), which is the characteristic property of non-abelian gauge theories. In the general case, given the non-commutative space $X$, i.e. the algebra $\mathcal{A}$, we have automatically the gauge groups $\mathcal{U}_n = \{u \in M_n(\mathcal{A}); uu^* = u^*u = 1\}$.[1] Moreover, the initial datum of non-commutative geometry, i.e. that of a Fredholm module (see Section 1) is exactly what is needed in order to define spinor fields on the space $X$ and the Dirac operator.

The gauge invariance principle allows us then to introduce 'gauge bosons' as operators in the Hilbert space. The principal difficulty is to write a local gauge invariant self-interaction for these gauge bosons. It is here that the principle of local interactions comes in. In order to reconstruct the self-interaction of the gauge bosons we shall use the following property of renormalizable theories:

(∗) For one-loop diagrams the calculated divergences occur additively, are *logarithmic* as functions of the 'cut-off' parameter $\Lambda$, and the coefficient of $\log \Lambda$ is a *local* function of the relevant fields which has *the same form as the Lagrangian* one started with.

This property, without being a characterization, is verified by renormalizable theories which appear in physics, provided one uses the gauge invariance to eliminate non-logarithmic divergences which do not have any physical significance. For a similar statement, see [4, p. 225]:

'There is a remarkable property characteristic of all divergent one loop integrals: ultraviolet infinities may be isolated in the form of additive components, the latter have the form of a polynomial in the derivatives of the $\delta$ function.'

This statement is less precise since it does not specify the degree of divergence, nor the relation between the Lagrangian and the counter-terms.

---

[1] For this it is important that $\mathcal{A}$ is an involutive algebra.

## 2. Essay on Physics and Non-commutative Geometry

We shall begin (Section 2) by establishing an invariant way of 'taking the coefficient of $\log \Lambda$', where $\Lambda$ is a 'cut-off'. For this it suffices to note that the contributions from one-loop diagrams are traces of operators and to use a result of J. Dixmier (1966), in which he constructs an operator trace (which we shall call the Dixmier trace) by taking the limit, in an ingenious fashion, of the sequence:

$$\frac{1}{\log N} \sum_0^N \mu_n(T),$$

where the $\mu_n$ are the eigenvalues of the positive operator $T$. This trace will be our principal tool, and it is remarkable that the motivation of J. Dixmier was to resolve a problem in pure mathematics: Is the only trace on the algebra of all bounded operators on a Hilbert space the standard one?

We shall show later (Section 4) how to invert statement 1. so as to define in a satisfactory way the self-interaction of gauge bosons, using only the Fredholm module given at the start.

But before that we prove (Section 3) purely mathematical results to show the power of the Dixmier trace. We shall show in particular that the logarithmic divergences, far from being an accident due to the 'commutativity' of the space $X$, are forced by the non-triviality of a certain Hochschild cohomology class, namely that of the character of the Fredholm module.

Finally, in Section 5, we shall discuss explicit models. The question which we answered there is the following: what $*$ algebra $\mathcal{A}$ and $K$-cycle $(\mathfrak{H}, D, \gamma)$ should we take so that our general action functional (Theorem 1 of Section 4) produces exactly the Weinberg–Salam model of electroweak unification. It is very satisfactory that this question has a positive answer. The picture that emerges is that of Fig. 2, namely a two-sheeted Euclidean space-time, where the distance of the two sheets is $\sim 10^{-16}$cm $= (M_W)^{-1}$, the inverse mass of the intermediate charged boson. The geometry of this double-sheeted space-time is specified by the conformal structure of Euclidean space-time and an auxiliary field $\Gamma$ which is discussed in detail in Section 5. This model is the simplest "Kaluza–Klein" theory, where the fibre is just a *2-point space* whose treatment requires the new ideas of non-commutative geometry. The *disconnectedness* of the fibre is crucial in that it allows for the *non-trivial* bundle with fibre $\mathbb{C}$ on one of the two copies and $\mathbb{C}^2$ on the other. The choice of connection is the symmetry breaking part of the Higgs sector.

I am grateful to Tsou S. T. for translating into English the original French version of this paper.

## 1. Gauge group, spinors and Dirac operators

If $\mathcal{A}$ is an involutive algebra and $n$ an integer, we obtain a group $\mathcal{U}_n$ by defining $\mathcal{U}_n = \{u \in M_n(\mathcal{A}), uu^* = u^*u = 1\}$. It is the unitary group of the algebra $M_n(\mathcal{A})$ of $n \times n$ matrices with coefficients in $\mathcal{A}$. It is easy to verify that if $\mathcal{A}$ denotes the algebra of functions on an ordinary space $X$ with certain smoothness conditions (for example $C^\infty$), then the corresponding group $\mathcal{U}_n$ is that of maps of $X$ into the compact group $U(n)$ with the same smoothness condition. In other words, $\mathcal{U}_n(C^\infty(X)) = C^\infty(X, U(n))$.

This way of constructing groups from algebras does not give all the groups, since the groups thus obtained have the remarkable property that their group law alone essentially contains the algebra structure of $\mathcal{A}$. An analogous phenomenon for the groups $GL_n(\mathcal{A})$ of invertible elements of $M_n(\mathcal{A})$ plays an essential role in algebraic $K$ theory (cf. J. Milnor, *Introduction to Algebraic K Theory*).

The complexified Lie algebra of $\mathcal{U}_n(\mathcal{A})$ can be identified with the Lie algebra of matrices $M_n(\mathcal{A})$, with the bracket $[a, b] = ab - ba$ and the product in $\mathcal{A}$ is determined[2] for example by the equation:

$$\left[ \begin{pmatrix} a & 0 \\ 0 & 0 \end{pmatrix}, \begin{pmatrix} 0 & b \\ 0 & 0 \end{pmatrix} \right] = \begin{pmatrix} 0 & ab \\ 0 & 0 \end{pmatrix}$$

for all $a, b \in \mathcal{A}$. (For an arbitrary Lie algebra $\mathfrak{l}$ one does not have a natural Lie algebra structure on $M_2(\mathfrak{l}) = M_2 \otimes \mathfrak{l}$. Moreover, the bracket can be written as $[a, b] = ab - ba$ by using the enveloping algebra, but it is not true that $ab \in \mathfrak{l}$.)

As in algebraic $K$ theory it is natural to keep the collection of all the groups $\mathcal{U}_n$ together with the obvious inclusions $\mathcal{U}_n \hookrightarrow \mathcal{U}_m$ ($m \geq n$) and $U(n) \subset \mathcal{U}_n$. Each unitary representation $\pi$ of the algebra $\mathcal{A}$ defines a representation $u \mapsto \pi(u)$ of $\mathcal{U}$ (and more generally of $\mathcal{U}_n$), but the resulting representations are very special since for example their tensor powers are no longer of the same form. We shall recognize from this feature the 1-particle space for fermions. In other words, 'the action of the gauge group of the second kind in the 1-particle space of fermions comes from a representation of the algebra $\mathcal{A}$.'

The Dirac operator, invented by Dirac to make Schrödinger's equation relativistic, had a considerable impact on the subsequent development of mathematics. We might suppose that we have already understood it from the description of the unitary representations of the Poincaré group, but it has, I believe, a more general significance

---

[2] Notice that we use the inclusion $M_n(\mathbb{C}) \subset M_n(\mathcal{A})$ in this formula, i.e. the inclusion of the gauge group of the first kind $U(n)$ in the gauge group $\mathcal{U}_n$ of the second kind.

as the generator of the $K$ homology $K_*(X)$ of a space $X$. Without going into a detailed description of this theory, let me point out that owing to the work of M. Atiyah in England, I. Singer and Brown-Douglas-Fillmore in the U.S., and Miscenko and Kasparov in Russia, it has played a crucial role in the topology of non-simply connected manifolds and in operator theory. What is remarkable is that the simplest way to formulate this theory uses as data Fredholm modules on the algebra $\mathcal{A}$ of functions on $X$, and that:

1. The commutativity of $\mathcal{A}$ is not at all important; in fact, quite the contrary (since it would be a pity not to make use of $M_n(\mathcal{A})$).

2. The *unbounded* formulation of the concept of Fredholm modules matches exactly the formulation of Dirac's theory of the electron.

We shall now give a more precise definition of an unbounded Fredholm module on an algebra $\mathcal{A}$, and define its dimension. To save on the terminology we shall use the term $K$-cycle.

**Definition 1** *A K-cycle on $\mathcal{A}$ is given by a unitary representation of $\mathcal{A}$ in a Hilbert space $\mathfrak{H}$ and an unbounded self-adjoint operator $D$ in $\mathfrak{H}$ such that*

1. *$[D, a]$ is bounded for all $a \in \mathcal{A}$*
2. *$(1 + D^2)^{-1}$ is compact.*

*(If $\mathcal{A}$ does not have an identity then we replace 2. by $a(1 + D^2)^{-1}$ compact for all $a \in \mathcal{A}$).*

Of course, the Dirac operator on a spin manifold defines a $K$-cycle on the algebra $\mathcal{A}$ of functions on the manifold. One has $\mathfrak{H} = L^2(M, S)$, the Hilbert space of square integrable sections of the spin bundle $S$ on the manifold $M$; the operator $D$ is the self-adjoint unbounded Dirac operator from $\mathfrak{H}$ to $\mathfrak{H}$, and the action of $\mathcal{A}$ in $\mathfrak{H}$ is by multiplication; if $f \in \mathcal{A}$ and $\xi \in \mathfrak{H}$ then $(f\xi)(x) = f(x)\xi(x)$ for all $x \in M$.

Whereas Gelfand's theory shows that given the algebra $\mathcal{A}$ one can recover the topological space $M$, it is not difficult to verify that given the Fredholm module $(\mathfrak{H}, D)$ above on $\mathcal{A}$ one can recover the Riemannian metric on $M$. In fact, if $P, Q \in M$ are two points in $M$, i.e. two characters $p(f) = f(P)$, $q(f) = f(Q)$, for all $f \in \mathcal{A}$ of the algebra $\mathcal{A}$, the *geodesic distance* $d(P, Q)$ of $P$ to $Q$ for the Riemannian structure is given by the formula:

$$d(P, Q) = \sup\{|p(f) - q(f)|;\ f \in \mathcal{A}, \|[D, f]\| \leq 1\},$$

where the norm $\|[D,f]\|$ is the operator norm in the Hilbert space $\mathfrak{H}$.

This means that no information is lost in replacing the local geometric data of the Riemannian structure on $M$ by the operator theoretic data of the unbounded Fredholm module or the $K$-cycle $(\mathfrak{H}, D)$. From our point of view, to analyse a non-commutative space in general, the strategy is the following:

1. By minimizing a suitable action functional $I$ to determine a $K$-cycle $(\mathfrak{H}, D)$ with a given $K$ homology class.

2. From a $K$-cycle $(\mathfrak{H}, D)$ to construct the analogue of the Lagrangian of gauge theories by using $D^{-1}$ as a fermion propagator, i.e. $D$ as a Dirac operator.

Step 2. consists in constructing also an action functional, whose argument is no longer $(\mathfrak{H}, D)$ as in 1. but gauge fields of the form $A = \sum a[D, b]$, where $a, b \in \mathcal{A}$.

It is to construct these two action functionals that it is necessary to understand better the notion first of the *dimension* of a $K$-cycle, then the *logarithmic divergences*. At the end of this section we shall make precise only the notion of dimension of a $K$-cycle. Logarithmic divergences will be the subject of Section 2.

Let $(\mathfrak{H}, D)$ be a $K$-cycle on $\mathcal{A}$. We consider the operator $D$ as a substitute for the Fourier transform and the orthonormal basis $(\xi_p)$ of eigenvectors of $D$, $D\xi_p = E(p)\xi_p$, as defining the countable 'momentum space'. We then measure the *dimension* of the module by the growth of the momentum space, i.e. more precisely:

**Definition 2** *Let $d \in [1, \infty)$ be a real number. A $K$-cycle $(\mathfrak{H}, D)$ on $\mathcal{A}$ is $d^+$ summable if the eigenvalues $E_n$ of $D$ arranged in increasing order satisfy:*

$$\sum_1^N E_n^{-1} = O(\sum_1^N n^{-1/d}).$$

For $d = 1$ this means that $\sum_1^N E_n^{-1} = O(\log N)$ and for $d > 1$ that $E_n^{-1} = O(n^{-1/d})$.

This condition gives an upper bound for the dimension of the $K$-cycle; we shall see later a cohomological non-triviality condition which gives a lower bound to this dimension. The interesting case (Section 3) is going to be the case where these two bounds coincide.

The Tauberian theorem of Ikehara shows that the $K$-cycle $(\mathfrak{H}, D)$ is $d^+$ summable whenever the function $T(s) = \text{Trace}(|D|^{-s})$ is holomorphic for $Re(s) > d$, continuous for $Re(s) \geq d$, $s \neq d$ and has a simple pole at $s = d$.

Let us mention another way of saying that the $K$-cycle $(\mathfrak{H}, D)$ is $d^+$ summable: $|D|^{-1} \in \mathcal{L}^{d+}(\mathfrak{H})$, where $\mathcal{L}^{d+}$ is the *two-sided ideal* of compact operators which is defined by the following condition on their eigenvalues:

$$T \in \mathcal{L}^{d+}(\mathfrak{H}) \iff \sum_0^N \mu_n(T) = O(\sum_0^N (1+n)^{-1/d}),$$

where $\mu_n(T)$ is the $n$th eigenvalue of $|T| = (T^*T)^{1/2}$.

These ideals $\mathcal{L}^{d+}$ play a crucial role in the work of D. Voiculescu [14] on the scattering theory of several variables and are the analogues of the spaces $L_w^p$, the weak $L^p$ of classical analysis. It is not clear a priori that they are even vector spaces, but this follows from the inequalities of Hermann Weyl:

$$\sum_0^N \mu_n(T_1 + T_2) \leq \sum_0^N \mu_n(T_1) + \sum_0^N \mu_n(T_2).$$

To end this section, let us give a construction of $2^+$ summable $K$-cycles on the algebra $\mathcal{A} = C_c^\infty(\mathbb{R}^n)$ of regular functions on $\mathbb{R}^n$, which is familiar in string theory: if $\Sigma$ is a Riemann surface and $\psi : \Sigma \to \mathbb{R}^n$ a regular map from $\Sigma$ to $\mathbb{R}^n$, we can form the composition of the $K$-cycle $(\mathfrak{H}, D)$ of spinors on $\Sigma$ and the homomorphism $\psi^* : \mathcal{A} \to C^\infty(\Sigma)$ which to $f \in \mathcal{A}$ assigns $f \circ \psi \in C^\infty(\Sigma)$. We obtain thus a $2^+$ summable $K$-cycle on $\mathcal{A}$ which characterizes the couple $(\Sigma, \psi)$, provided $\psi$ is an immersion.

## 2. Dixmier trace and logarithmic divergences

The divergences which appear in perturbative field theory, as contributions to the perturbation series of one-loop Feynman diagrams, are typically logarithmic. This means that if in calculating the integral in momentum space we neglect energies $E > \Lambda$, i.e. we restrict the integral to $|p| \leq \Lambda$ (where $\Lambda$ is the cutoff parameter), we obtain a finite quantity which, however, diverges as $O(\log \Lambda)$ when $\Lambda \to \infty$. Moreover, by confining oneself to compact spaces, which is equivalent to discretizing momentum space to avoid problems of infrared divergences, one can easily write the contribution of one-loop Feynman diagrams as the trace: Trace($T$) of an operator $T$ in the one-particle space of free fields, constructed from Green's functions or propagators for free fields. Thus, typically, Trace($T$)= $\sum \lambda_n(T)$ diverges logarithmically as a function of the cutoff $\Lambda$, and it is not difficult to see that if one counts only the first $N$ eigenvalues of $|T|$ one has the equivalence $N \sim \Lambda^d$ in terms of a cutoff of energy $E \leq \Lambda$.

One is thus confronted with the existence of numerous operators $T$ whose ordinary trace diverges logarithmically but for which one would wish to define properly the coefficient of $\log N \sim d \log \Lambda$ in the divergent term.

It is precisely this problem which J. Dixmier resolved in 1966 in a note in the *Comptes Rendus* [9], the motivation for which was the following mathematical problem: 'Is the ordinary trace the only trace on the algebra of bounded operators in a Hilbert space?'

Instead of rewriting Dixmier's article, which is both short and precise, I prefer to try and show the relation between his solution and the ideas of the 'renormalization group fixed point' in physics applied in a *linear* context and hence very simple.

The naive idea would be to define a functional $\phi$ on the positive operators $T \in \mathcal{L}^{1+}(\mathfrak{H})$ by the equation:

$$\phi(T) = \lim_{N \to \infty} \frac{1}{\log N} \sum_0^N \mu_n(T),$$

where the $\mu_n$ are the eigenvalues arranged in decreasing order. By hypothesis $T \in \mathcal{L}^{1+}(\mathfrak{H})$, so that the series $(1/\log N) \sum_0^N \mu_n(T) = a_N$ is *bounded*.

The problem is that this series does not converge in general, and also that the hypothesis that it does converge does not define a two-sided ideal. All that one knows is that the series $(a_N)$ does not change if one replaces $T$ by a unitarily equivalent operator: $a_N(UTU^*) = a_N(T)$ (since $\mu_n(UTU^*) = \mu_n(T)$) and that one has the inequality:

$$(*) \quad \sum_0^N \mu_n(T_1 + T_2) \leq \sum_0^N \mu_n(T_1) + \sum_0^N \mu_n(T_2) \leq \sum_0^{2N} \mu_n(T_1 + T_2)$$

i.e.

$$a_N(T_1 + T_2) \leq a_N(T_1) + a_N(T_2) \leq a_{2N}(T_1 + T_2) \times \frac{\log 2N}{\log N}.$$

It is then sufficient, in order to obtain a trace, to have a limiting procedure, denoted $\lim_w$, which to every bounded sequence $(a_N)$ of real numbers associates a real number $\lim_w(a_N)$ such that:

$$(**) \quad \lim_w(a_{2N}) = \lim_w(a_N); \quad a_N \geq 0 \Longrightarrow \lim_w(a_N) \geq 0.$$

Of course, we want this procedure to be non-trivial, say by imposing the condition $\lim_w(1) = 1$.

Now the condition $(**)$ means exactly that the functional $\psi = \lim_w$ is *scale invariant*, i.e. it is a fixed point of the action by the group of scale changes $\Lambda \mapsto \lambda\Lambda$, i.e. of the dilatations of momentum space. J. Dixmier uses the amenability of the group of triangular matrices $\begin{pmatrix} a & b \\ 0 & a^{-1} \end{pmatrix}$ to construct such a limiting procedure (also (**) is valid only provided $a_{N+1} - a_N \to 0$ when $N \to \infty$). In his paper he uses a more general series than the series $\log N$ above, but it is clear that one cannot replace $\log N$ by $N^\alpha > 0$, since $(2N)^\alpha/N^\alpha$ does not tend to 1 as $N \to \infty$.

Another more subtle point concerning the Dixmier trace $\text{Tr}_w$, which for the operators $T \geq 0$, $T \in \mathcal{L}^{1+}$ is given by:

$$\text{Tr}_w(T) = \lim_w \frac{1}{\log N} \sum_0^N \mu_n(T),$$

is its dependence on the limiting procedure which exists by virtue of the amenability of a group. One might thus think that such a procedure has as precarious an existence as that of a (nonprincipal) ultrafilter on $\mathbb{N}$, i.e. a non-measurable character of the group $(\mathbb{Z}/2)^{\mathbb{N}}$. In fact this is not at all the case, since one does not impose the condition

$$\lim_w(a_N b_N) = \lim_w(a_N)\lim_w(b_N)$$

which is incompatible with (**). Moreover, as (**) is a *linear* condition on the functional $\lim_w = \psi$ a remarkable result of G. Mokobodski [12] shows the existence (assuming the continuum hypothesis) of a procedure $\lim_w$ which, besides being *universally measurable* as a function of the sequence $(a_N)$, in addition *satisfies the formula*:

$$\lim_w(\int a_N(\alpha)d\mu(\alpha))) = \int d\mu(\alpha)\lim_w(a_N(\alpha)),$$

where $(a_N(\alpha))$ is a measurable family of bounded sequences indexed by an arbitrary probability space $(X, \mu)$.

## 3. Cohomological dimension of a $K$-cycle and the necessity of logarithmic divergences

The starting point of cyclic cohomology is the possibility, given a $K$-cycle $(\mathfrak{H}, D)$ on an algebra $\mathcal{A}$ (and a $\mathbb{Z}/2$ grading $\gamma$ in $\mathfrak{H}$ such that $\gamma a = a\gamma$ for $a \in \mathcal{A}$, $\gamma D = -D\gamma$), of imitating the differential and integral calculus of differential forms on a manifold in the following manner. The sign $F = D|D|^{-1}$ of the operator $D$ (supposed to be invertible for simplicity) enables us to define an operator-valued differential $da = i[F, a]$ for $a \in \mathcal{A}$, and the property $F^2 = 1$ shows that $d(da) = 0$ (where for $X \in \mathcal{L}(\mathfrak{H}), X\gamma = -\gamma X$, we put $dX = i(FX + XF)$ instead of $FX - XF$). We get then the complex

$$\mathcal{A} \xrightarrow{d} \Omega^1(\mathcal{A}) \xrightarrow{d} \Omega^2(\mathcal{A}) \to \cdots,$$

where $\Omega^k(\mathcal{A})$ is the vector space of linear combinations of operators of the form $\omega = a^0 da^1 \ldots da^k$, $a^j \in \mathcal{A}$.

The composition of operators in $\mathfrak{H}$ allows one to define the product of quantized forms, for $\omega_1 \in \Omega^{k_1}$, $\omega_2 \in \Omega^{k_2}$, one has $\omega_1 \omega_2 \in \Omega^{k_1+k_2}$, $d(\omega_1 \omega_2) = (d\omega_1)\omega_2 + (-1)^{\delta\omega_1} \omega_1 d\omega_2$. If the $K$-cycle $(\mathfrak{H}, D)$ is $d^+$ summable (cf. above), the quantized forms $\omega \in \Omega^k(\mathcal{A})$ have the following property: $\sum \mu_n(\omega)^p < \infty$ for all $p > d/k$, where the $\mu_n(\omega)$ are the eigenvalues of the operator $\omega$. (In more learned terms, one has $\Omega^k \subset \mathcal{L}^p$, the Schatten ideal of $p$ summable operators, for all $p > d/k$.)

This property is sufficient for defining the analogue of the integral for the quantized forms $\omega \in \Omega^d(\mathcal{A})$ (assuming $d$ an *even integer*) by the equation:

$$\int \omega = \mathrm{Trace}(\gamma \omega).$$

This formula does not make any sense as such, since for $\omega \in \Omega^d$ one has $\sum \mu_n(\omega)^p < \infty$ only for $p > 1$, but one can regularize it very simply by replacing $\mathrm{Trace}(\gamma\omega)$ by $\mathrm{Tr}_s(\omega) = \frac{1}{2}\mathrm{Trace}(\gamma F(F\omega + \omega F))$. As $F\omega + \omega F \in \Omega^{d+1}$ there is no problem whatsoever. One then proves the analogue of Stokes' law:

$$\int d\omega = 0, \quad \forall \omega \in \Omega^d(\mathcal{A}),$$

as well as the following property of commutativity under the integral sign:

$$\int \omega_1 \omega_2 = (-1)^{\delta_1 \delta_2} \int \omega_2 \omega_1,$$

where $\omega_1$ and $\omega_2$ are of degree $\delta_1, \delta_2$ with $\delta_1 + \delta_2 = d$.

This formalism, which leads directly to cyclic cohomology, allows one to extend successfully in highly non-commutative cases the methods in *differential topology* (see for example the application described in [8] and in the article by E. Effros in the *Mathematical Intelligencer*).

The notion of a *cyclic cocycle* $\tau$ on an algebra $\mathcal{A}$ is analogous to that of a trace. A trace $\tau_0$ on $\mathcal{A}$ is a linear functional on $\mathcal{A}$ which satisfies the following property:

$$(*)_0 \quad \tau_0(x^0 x^1) - \tau_0(x^1 x^0) = 0 \quad \forall x^0, x^1 \in \mathcal{A}.$$

A cyclic $n$-cocycle $\tau_n$ on the algebra $\mathcal{A}$ is an $n+1$ linear functional on $\mathcal{A}$ satisfying:

$$(*)_n \quad \begin{aligned} &\tau(x^0 x^1, \ldots, x^{n+1}) - \tau(x^0, x^1 x^2, \ldots, x^{n+1}) + \cdots \\ &+ (-1)^j \tau(x^0, \ldots, x^j x^{j+1}, \ldots, x^{n+1}) + \cdots \\ &+ (-1)^{n+1} \tau(x^{n+1} x^0, \ldots, x^n) = 0 \end{aligned}$$

$$(\lambda)_n \quad \tau(x^1, \ldots, x^n, x^0) = (-1)^n \tau(x^0, \ldots, x^n),$$

for all $x^j \in \mathcal{A}$.

The homological information given by a $K$-cycle $(\mathfrak{H}, D)$ on $\mathcal{A}$ as above is captured by its *character* which is the following cyclic $d$-cocycle:

$$\begin{aligned} \tau(x^0, x^1, \ldots, x^d) &= \int x^0 dx^1 \ldots dx^d \\ &= (-1)^{d/2} \mathrm{Tr}_s(x^0 [F, x^1] \ldots [F, x^d]). \end{aligned}$$

If $\mathcal{A}$ is the algebra of functions on the manifold $M$, a closed de Rham current $\mathbf{C}$ of dimension $n \in \{0, 1, \ldots, \dim M\}$ defines a cyclic cocycle $\tau_\mathbf{C}$ on $\mathcal{A}$ by the equation:

$$\tau_\mathbf{C}(f^0, \ldots, f^n) = \langle \mathbf{C}, f^0 df^1 \wedge \cdots \wedge df^n \rangle, \quad \forall f^j \in \mathcal{A},$$

and the construction of the Chern character:

$$\mathrm{ch} : K^*(M) \to H^*(M, \mathbb{C}),$$

which, using connections and curvatures, assigns to each complex vector bundle $E$ on $M$ a cohomology class $\mathrm{ch}(E)$ with complex coefficients, is a particular case of the following result, which no longer requires the commutativity of $\mathcal{A}$:

**Lemma 1** *Let $\mathcal{A}$ be an algebra and $\tau_{2m}$ a cyclic cocycle on $\mathcal{A}$. The following equation defines an additive map from $K_0(\mathcal{A})$ to $\mathbb{C}$:*

$$\langle e, \tau_{2m} \rangle = \tau_{2m}(e, \ldots, e) \quad \forall e \in \mathrm{Proj}(M_k(\mathcal{A})).$$

We should point out here that the concept of a 'complex vector bundle' on a manifold $M$ is identical to that of a *finitely generated projective module* $\mathcal{E}$ on the algebra of functions on $M$. The module associated to a bundle $E$ is that of the sections of $E$. Moreover, every finitely generated projective module $\mathcal{E}$ on an algebra $\mathcal{A}$ (with an identity) is of the form $\mathcal{E} = e\mathcal{A}^k$, where $e$ is a projection (i.e. $e^2 = e$), $e \in M_k(\mathcal{A})$.

In the above statement, one should equally extend the cocycle $\tau$ to the algebra of matrices $M_k(\mathcal{A})$ on $\mathcal{A}$, which can be done by the equation:

$$\tau(\mu^0 \otimes a^0, \ldots, \mu^{2m} \otimes a^{2m}) = \mathrm{trace}(\mu^0 \ldots \mu^{2m}) \tau(a^0, \ldots, a^{2m}),$$

with $\mu^j \in M_k(\mathbb{C})$, $a^j \in \mathcal{A}$.

In the particular case where $\mathcal{A}$ is the algebra of functions on a manifold $M$ and $\tau = \tau_\mathbf{C}$ is the associated closed current $\mathbf{C}$ we obtain the equation:

$$\langle e, \tau_\mathbf{C} \rangle = \langle \mathrm{ch}(e), \mathbf{C} \rangle,$$

which determines uniquely the Chern character of the bundle $E$ associated to the projection $e \in M_k(\mathcal{A})$.

The interest of this pairing in the case where $\mathcal{A}$ is no longer commutative is explained in detail in conjunction with the quantum Hall effect (cf. the work of J. Bellissard [3]).

The integrality property which plays a crucial role in the quantum Hall effect is the particular case of the integrality of the pairing above for the character of a Fredholm module. In fact we have:

**Proposition 1** *Let $\mathcal{A}$ be an algebra, $(\mathfrak{H}, D, \gamma)$ a $d^+$ summable K-cycle on $\mathcal{A}$ where $d = 2m$ is an even integer, and $\tau_{2m}$ the character of $(\mathfrak{H}, D, \gamma)$. Then one has for all $e \in \text{Proj}(M_k(\mathcal{A}))$:*

$$\langle \tau_{2m}, e \rangle = \text{Ind}(D_e^+) \in \mathbb{Z},$$

*where $D_e^+$ is the Fredholm operator obtained by restricting the operator $D^+ = (\frac{1-\gamma}{2})D(\frac{1+\gamma}{2})$ to the image of $e$.*

In the definition of the character $\tau_d$ of a $K$-cycle, the only condition on the even integer $d$ is that the $K$-cycle is $d^+$ summable, i.e. $\sum_1^N \lambda_n^{-1} = O(\sum_1^N n^{-1/d})$ where the $\lambda_n$ are the eigenvalues of $|D|^{-1}$. It is clear that if this condition is satisfied with $d = 2m$, then it is also true for $d = 2m + 2q$, $q \in \mathbb{N}$. There are then two natural questions to ask:

1. How does one relate the cyclic cocycles $\tau_{2m+2q}$ to $\tau_{2m}$?

2. How does one decide which $d$ is the smallest possible?

The answer to 1. is very simple and is at the origin of the periodicity operator $S$ in cyclic cohomology.

Let us begin by eliminating those uninteresting $n$-cocyles which are obtained by the relation $\tau = b\psi$, where $\psi$ is an $n$-linear functional on $\mathcal{A}$ satisfying the following condition:

$$(\lambda)_{n-1} \quad \psi(x^1, x^2, \ldots, x^{n-1}, x^0) = (-1)^{n-1}\psi(x^0, \ldots, x^n), \quad \forall x^i \in \mathcal{A},$$

and where the coboundary operator is given by the equation:

$$\begin{aligned}(b\psi)(x^0, \ldots, x^n) &= \sum (-1)^j \psi(x^0, \ldots, x^j x^{j+1}, \ldots, x^n) \\ &\quad + (-1)^n \psi(x^n x^0, \ldots, x^{n-1}).\end{aligned}$$

We have $b^2 = 0$, i.e. $b(b\psi) = 0$ for all $\psi$, and moreover if $\psi$ satisfies $(\lambda)^\bullet_{n-1}$ then $b\psi$ satisfies $(\lambda)_n$. Given the arbitrariness of the $\psi$ the cyclic cocycles $b\psi$ has absolutely no cohomological content and we eliminate them by introducing the cyclic cohomology groups: $H_\lambda^n(\mathcal{A}) = HC^n(\mathcal{A})$, quotient of the space of cyclic $n$-cocycles by the coboundaries of cyclic cochains.

The link between the characters $\tau_d$ and $\tau_{d+2k}$ of $d^+$ summable $K$-cycle is then given by the following result:

**Theorem 1** *If $(\mathfrak{H}, D)$ is a $d^+$ summable $K$-cycle, one has, for all $k$, $\tau_{d+2k} = S^k \tau_d \in H_\lambda^{d+2k}(\mathcal{A})$, where $S$ is the operator defined in general by:*

$$S\phi = \phi \otimes \sigma \in H_\lambda^{n+2}(\mathcal{A}) \quad \forall \phi \in H_\lambda^n(\mathcal{A}),$$

*where $\sigma$ denotes the following generator of the ring $H_\lambda^*(\mathbb{C})$ :$\sigma(1,1,1) = 1$.*

One uses the identification $\mathcal{A} \otimes \mathbb{C} = \mathcal{A}$, as well as the existence of a natural tensor product $H_\lambda^n(\mathcal{A}) \otimes H_\lambda^m(\mathcal{B}) = H_\lambda^{n+m}(\mathcal{A} \otimes \mathcal{B})$.

The above theorem gives a complete answer to the question 1. In order to answer question 2. it is then sufficient to characterize the image $\text{Im}(S) \subset H_\lambda^n(\mathcal{A})$ of the periodicity operator; in fact if the chosen $d$ is not optimal and can be replaced by $d - 2$ then one has $\tau_d \in \text{Im} S$ according to the theorem. The second crucial result of cyclic cohomology is the following characterization of the image of $S$:

**Theorem 2** *In order for $\phi \in H_\lambda^n(\mathcal{A})$ to lie in $\text{Im}(S)$ it is necessary and sufficient that there exists an $n$ linear functional $\psi$ on $\mathcal{A}$ (not necessarily satisfying $(\lambda)_{n-1}$) such that $b\psi = \phi$.*

Of course, if $\psi$ is an arbitrary $n$-linear form on $\mathcal{A}$, $b\psi$ does not satisfy $(\lambda)_n$ in general and is hence not a cyclic cocycle. The cohomology of the complex of arbitrary $(n+1)$-linear forms is identified with the Hochschild cohomology $H^n(\mathcal{A}, \mathcal{A}^*)$ of $\mathcal{A}$ with values in the bimodule $\mathcal{A}^*$ of linear forms on $\mathcal{A}$. Theorem 2 means that one has the following exact sequence:

$$H_\lambda^{n-2}(\mathcal{A}) \xrightarrow{S} H_\lambda^n(\mathcal{A}) \xrightarrow{I} H^n(\mathcal{A}, \mathcal{A}^*),$$

where $I$ is the map which 'forgets' the cyclicity condition $(\lambda)_n$. One then has:

**Corollary** *Let $(\mathfrak{H}, D, \gamma)$ be a $d^+$ summable $K$-cycle on $\mathcal{A}$ and suppose that the Hochschild class $I(\tau_d)$ of the character is non-zero, then the integer $d$ is optimal. If $I(\tau_d) = 0$, then there exists a $d - 2$ cyclic cocycle $\phi$ on $\mathcal{A}$ such that $\tau_d = S\phi$.*

The Hochschild cohomology $H^k(\mathcal{A}, \mathcal{A}^*)$ is directly calculable as a result of the tools developed (cf. [8]) for the needs of homological algebra. For example, if $\mathcal{A}$ is the algebra $C^\infty(M)$ of $C^\infty$ functions on a manifold $M$, and if we require the cochains we consider to be continuous, then we obtain the relation $H^k(\mathcal{A}, \mathcal{A}^*) = \Omega_k$, where $\Omega_k$ is the space of *de Rham currents* of dimension $k$ on $M$, i.e. continuous linear forms on the space $C^\infty(M, \Lambda^k T^*)$ of differential forms of degree $k$ on $M$. The de Rham boundary operator is in fact a special case of the operator $B : H^k(\mathcal{A}, \mathcal{A}^*) \to H_\lambda^{k-1}(\mathcal{A})$, and one obtains then the principal tool for calculating cyclic cohomology:

**Theorem 3** *For all algebras $\mathcal{A}$ one has the following long exact sequence:*

$$\xrightarrow{B} H_\lambda^{n-2}(\mathcal{A}) \xrightarrow{S} H_\lambda^n(\mathcal{A}) \xrightarrow{I} H^n(\mathcal{A}, \mathcal{A}^*) \xrightarrow{B}$$
$$H_\lambda^{n-1}(\mathcal{A}) \xrightarrow{S} H_\lambda^{n+1}(\mathcal{A}) \xrightarrow{I} H^{n+1}(\mathcal{A}, \mathcal{A}^*) \xrightarrow{B}$$

One has thus a complete answer to questions 1. and 2. in determining the image of $S$ by means of Hochschild cohomology and the kernel of $S$ by means of the image of $B$. It remains to calculate, given a $d^+$ summable $K$-cycle, the Hochschild class $I(\tau_d) \in H^d(\mathcal{A}, \mathcal{A}^*)$, i.e. the obstruction to decrease $d$ by two. *It is on this crucial point that the Dixmier trace $\mathrm{Tr}_w$ defined in Section 2 comes into play.* In fact one has the following very general result:

**Theorem 4** *Let $(\mathfrak{H}, D, \gamma)$ be a $d^+$ summable $K$-cycle on the algebra $\mathcal{A}$, $d = 2m$ an even integer. For all $w$, the following formula defines a Hochschild cocycle on $\mathcal{A}$:*

$$\psi_w(a^0, \ldots, a^d) = Tr_w(\gamma a^0 [D, a^1] \cdots [D, a^d] |D|^{-d}).$$

*Moreover, if the Hochschild cohomology of $\mathcal{A}$ is the dual of the Hochschild homology, each $\psi_w$ is equal to $I(\tau_d)$.*

Thus, the obstruction to decreasing $d$ by two (cf. Corollary) is measured by the value of the Dixmier trace, and in particular *logarithmic divergences* whose coefficient is $\mathrm{Tr}_w$ are *imposed by the non-triviality of $I(\tau_d)$*. We see from this theorem that far from being the fruit of coincidence or the corollary of the commutativity of $\mathcal{A}$, the logarithmic divergences are a general phenomenon, necessitated by the non-triviality of a Hochschild cocycle: $I(\tau_d)$. By virtue of Theorem 4 it is very easy to calculate this cocycle in examples. In particular, the Dixmier trace of a pseudo-differential operator $P$ of order $-n$ on a manifold $M$ of dimension $n$ coincides with the Adler-Manin-Wodzicki

[6] residue and can be calculated using the principal symbol $\sigma_P$ of $P$ by the formula:

$$\mathrm{Tr}_w(P) = \frac{1}{n}\frac{1}{(2\pi)^n}\int_{S^*M} \mathrm{trace}(\sigma_P)d\mu,$$

where $S^*M$ denotes the bundle of half-lines in the cotangent bundle of $M$ equipped with its canonical contact structure. One obtains then the computation of $I(\tau_d)$ for the $K$-cycles associated to elliptic operators of order 1, at the end of Section 1. If for example $\mathcal{A} = C^\infty(\mathbb{R}^n)$ and $(\mathfrak{H}, D, \gamma)$ is the $K$-cycle associated to a map $\psi$ of a Riemann surface $\Sigma$ to $\mathbb{R}^d$, one obtains $I(\tau_d) =$ the de Rham current of integration against the cycle $\psi(\Sigma)$. This current determines, and is determined by, at least when $\psi$ is an embedding, the image of $\Sigma$ in $\mathbb{R}^d$.

## 4. Positivity in Cyclic Cohomology and the Yang-Mills Action Functional

Theorem 4 of the last section is a basic result which allows one to express the Hochschild cohomology class of a $d^+$ summable $K$-cycle from the Dixmier trace of suitable products of commutators. But the Dixmier trace enjoys the fundamental property of being *positive*:

$$\mathrm{Tr}_w(A^*A) \geq 0, \quad \text{for any operator } A.$$

This shows that the Hochschild class of a $d^+$ summable $K$-cycle $(\mathfrak{H}, D, \gamma)$ has in fact a *positive* representative given by the following Hochschild cocycle:

$$(*) \quad \phi_w(a^0, \cdots, a^d) = \mathrm{Tr}_w((1+\gamma)a^0[D, a^1]\cdots[D, a^d]D^{-d}),$$

for any $a^0, \cdots, a^d \in \mathcal{A}$.

By definition (cf. [5]) a Hochschild cocycle $\phi$ on $\mathcal{A}$ of dimension $2m = d$ is *positive* iff the following equality defines a positive inner product on the linear space $\mathcal{A}^{\otimes(m+1)}$:

$$\langle a^0 \otimes a^1 \cdots \otimes a^m, b^0 \otimes b^1 \cdots \otimes b^m\rangle = \phi(b^{0*}a^0, a^1, \cdots, a^m, b^{m*}, \cdots, b^{1*}),$$

for any $a^j, b^j \in \mathcal{A}$.

For $m = 0$ this is the familiar definition of a positive form (a trace if $b\phi = 0$) on $\mathcal{A}$.

In general the positive Hochschild cocycles form a convex cone $Z_+^d \subset Z^d = Z^d(\mathcal{A}, \mathcal{A}^*)$ in the linear space $Z^d$ of Hochschild cocycles. To get familiar with this notion of positivity we consider the special case of the $*$-algebra of smooth functions $\mathcal{A} = C^\infty(M)$ on a compact manifold $M$ (and we restrict to *continuous* multilinear forms on $\mathcal{A}$).

For $d = 0$, the space $Z^0 = Z^0(\mathcal{A}, \mathcal{A}^*)$ is the space of 0-dimensional currents on $M$ and the cone $Z_+^0$ is the cone of *positive measures* in the usual sense.

For $d = 2$, a Hochschild cohomology *class* $C$ is characterised by a 2-dimensional de Rham current $\mathbf{C}$ which is obtained from any cocycle $\phi \in C \subset Z^2$ by the equality:

$$\langle \mathbf{C}, f^0 df^1 \wedge df^2 \rangle = \frac{1}{2}(\phi(f^0, f^1, f^2) - \phi(f^0, f^2, f^1)), \ \forall f^0, f^1, f^2 \in \mathcal{A}.$$

There is a unique element $\phi_C \in C$ which is skew symmetric in the arguments $f^1, f^2$, and it is given by:

$$\phi_C(f^0, f^1, f^2) = \langle \mathbf{C}, f^0 df^1 \wedge df^2 \rangle, \ \forall f^j \in \mathcal{A}.$$

This map $C \mapsto \phi_C$ is the natural cross-section that one usually uses to identify de Rham currents with cocycles rather than cohomology classes. It does however have a drawback: $\phi_C$ can never be positive. Indeed if $\phi_C$ is positive for each $f$ the following linear functional on $\mathcal{A}$ is a positive measure: $L_f(g) = \langle \mathbf{C}, g\, df \wedge d\bar{f} \rangle$. Thus as $L_{\bar{f}} = -L_f$ one gets $L_f = 0$ for any $f$ and $\mathbf{C}=0$. This shows that if one sticks to this canonical representative $\phi_C$ the notion of positivity remains hidden.

Let us now give examples of 2-dimensional Hochschild classes $C$ which have a *positive* representative: $\phi \in C \cap Z_+^2$. Thus, let $\Sigma$ be an *oriented* 2-dimensional compact submanifold of $M$, $\Sigma \subset M$ and let $\mathbf{C}$ be the current on $M$ which is $\sqrt{-1}$ times the fundamental class of $\Sigma$:

$$\langle \mathbf{C}, f^0 df^1 \wedge df^2 \rangle = i \int_\Sigma f^0 df^1 \wedge df^2, \quad \forall f^j \in \mathcal{A}.$$

We let $C$ be the Hochschild cohomology *class* $\in H^2(\mathcal{A}, \mathcal{A}^*)$ associated to the current $\mathbf{C}$.

Now let us *choose* a conformal structure $g$ on $\Sigma$, the following equality defines a *positive* cocycle which belongs to the class $C$:

$$\phi^g(f^0, f^1, f^2) = 2i \int_\Sigma f^0\, \partial f^1 \wedge \bar{\partial} f^2.$$

One checks indeed that:

$$\frac{1}{2}(\phi^g(f^0, f^1, f^2) - \phi^g(f^0, f^2, f^1)) = i \int_\Sigma f^0 df^1 \wedge df^2,$$

and that $\phi^g$ is positive, as is the natural inner product on forms of type (1,0) on the complex manifold $\Sigma$.

By construction $\phi^g$ depends upon the choice of $g$, and in fact the complex structure of $\Sigma$ is easily recovered from the cocycle $\phi^g$ from

## 2. Essay on Physics and Non-commutative Geometry

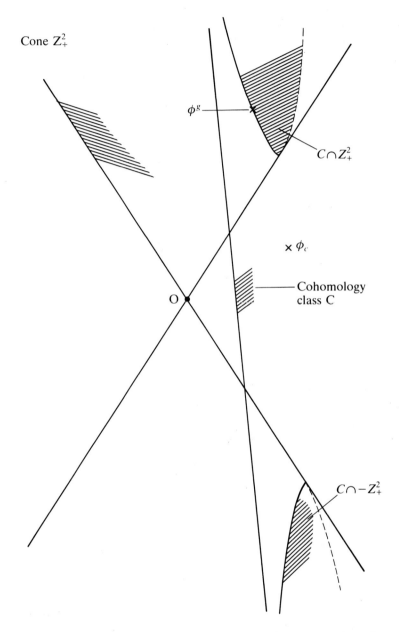

Fig. 1: Different conformal structures $g$ give different $\phi^g \in C \cap Z_+^2$

which one can reconstruct the $\mathcal{A}$-bimodule of $L^2$ forms of type (1,0) and the differential $\partial$. The situation is summarized in Figure 1.

One can show that for any non-degenerate *positive* element $G = \sum(dx^\mu)^* g_{\mu\nu} dx^\nu$ of the dual cone of $Z_+^2$, where $x^\mu \in \mathcal{A}$ and $(g_{\mu\nu})$ is a positive matrix of elements of $\mathcal{A}$, the following analogue of the Polyakov action reaches its minimum in $Z_+^2$ at the unique point $\phi^g$, where $g$ is the Riemannian structure on $\Sigma$ induced from the metric $\sum_{\mu,\nu} g_{\mu\nu}(dx^\mu)^* dx^\nu$ on $M$:

$$I(\phi) = \sum_{\mu,\nu} \phi(g_{\mu,\nu}, x^\nu, (x^\mu)^*).$$

For $d = 4$ as for $d = 2$ a Hochschild *class* $C$ is characterized by a 4-dimensional de Rham current $\mathbf{C}$ given by

$$\langle \mathbf{C}, f^0 df^1 \wedge \ldots \wedge df^4 \rangle = \frac{1}{4!} \sum (-1)^\sigma \phi(f^0, f^{\sigma(1)}, \ldots, f^{\sigma(4)}),$$

$\forall f^j \in \mathcal{A}$, for any cocycle $\phi \in C$.

Also, as for $d = 2$, the canonical cocycle $\phi_C$, which is the only skew symmetric element of the class $C$, can never be positive.

To give examples of 4-dimensional Hochschild classes $C$ which have positive representatives: $\phi \in C \cap Z_+^4$, we let $\Sigma$ be an oriented compact 4-dimensional submanifold of $M$ and use the following lemma:

**Lemma 1** *1) Let $g$ be a Riemannian metric on $\Sigma$ (assumed to be spin) and $D$ the corresponding Dirac operator in $\mathfrak{H} = L^2(\Sigma, S)$ with $\mathbb{Z}/2$ grading $\gamma$. Then the class of the following positive 4-Hochschild cocycle on $C^\infty(\Sigma)$ is the fundamental class of $\Sigma$:*

$$\phi_w(f^0, \ldots, f^4) = 8\pi^2 \mathrm{Tr}_w((1+\gamma) f^0 [D, f^1] \cdots [D, f^4] D^{-4}).$$

*2) The cocycle $\phi_w$ is independent of $w$; it depends only upon the conformal structure of $\Sigma$ and is equal to:*

$$\frac{1}{4} \int_\Sigma \mathrm{tr}((1+\gamma) f^0 df^1 . df^2 \ldots . df^4) dv = \phi^g(f^0, \ldots, f^4).$$

In the last expression the trace tr is the natural trace in the Clifford algebra at every point $x \in \Sigma$, the differentials $df^j$ are considered as sections of the Clifford algebra bundle and $dv$ is the Riemannian volume element on $\Sigma$.

The result easily follows from [6] Theorem 1, p. 674.

We shall now show how to reconstruct the Yang-Mills action functional on vector potentials from the Dirac operator $D$ using the positive Hochschild cocycle of the above lemma. The standard Lagrangian

## 2. Essay on Physics and Non-commutative Geometry

will then be easy to write down. The only difficulty is to make sure that our construction does not make use of the commutativity of the algebra $\mathcal{A}$ or of special properties of the Dirac operator other than giving a $K$-cycle over $\mathcal{A}$, and that we get in general a *positive quartic and gauge invariant* action on the affine space of vector potentials. To make sure of the nontriviality of this functional we shall extend to our context the usual inequality between the Chern class $c_2(E)$ of a vector bundle and the minimum of the Yang-Mills action on connections on $E$ (cf. Theorem 2). The notion of hermitian vector bundle extends easily to our context (cf. [7] for instance). We thus assume to be given a finite projective right module $\mathcal{E}$ over $\mathcal{A}$ together with a hermitian structure, i.e. an inner product $\langle \xi, \eta \rangle \in \mathcal{A}$ for $\xi, \eta \in \mathcal{E}$ satisfying suitable conditions (cf. [7]). The reader unfamiliar with this can take $\mathcal{E} = \mathcal{A}$ with module structure given by right multiplication by elements of $\mathcal{A}$ and inner product:

$$\langle a, b \rangle = a^* b, \quad \forall a, b \in \mathcal{E} = \mathcal{A}.$$

The vector potentials or compatible connections on $\mathcal{E}$ will be obtained as equivalence classes of non-local objects which we shall call universal vector potentials.

Let $\Omega^1(\mathcal{A})$ be the subspace of $\mathcal{A} \otimes \mathcal{A}$ given by the kernel of the multiplication $m : \mathcal{A} \otimes \mathcal{A} \to \mathcal{A}$, $m(a \otimes b) = ab \in \mathcal{A}$. It is a bimodule over $\mathcal{A}$ as a submodule of $\mathcal{A} \otimes \mathcal{A}$ with

$$a_1(a \otimes b)a_2 = a_1 a \otimes b a_2, \quad \text{for } a_1, a_2, a, b \in \mathcal{A}.$$

One defines a derivation $d$ of $\mathcal{A}$ in $\Omega^1(\mathcal{A})$ by:

$$da = 1 \otimes a - a \otimes 1. \quad \text{(cf. [11])}$$

**Definition 1** *Let $\mathcal{E}$ be a hermitian finite projective module over $\mathcal{A}$. Then a universal compatible connection on $\mathcal{E}$ is a linear map $\nabla$ of $\mathcal{E}$ to $\mathcal{E} \otimes_{\mathcal{A}} \Omega^1$ such that:*
*a) $\nabla(\xi a) = (\nabla \xi)a + \xi \otimes da \quad \forall \xi \in \mathcal{E}, \, a \in \mathcal{A}$*
*b) $\langle \nabla \xi, \eta \rangle + \langle \xi, \nabla \eta \rangle = d \langle \xi, \eta \rangle \quad \forall \xi, \eta \in \mathcal{E}.$*

Such connections always exist ([7]). If for instance we take $\mathcal{E} = e\mathcal{A}^n$, where $e$ is a selfadjoint idempotent, $e \in \text{Proj}(M_n(\mathcal{A}))$, then $\nabla \xi = ed\xi$ defines a universal compatible connection on $\mathcal{E}$ for the hermitian structure obtained by restriction of the obvious one on $\mathcal{A}^n$ : $\langle \xi, \eta \rangle = \sum \xi_i^* \eta_i \in \mathcal{A}$.

By construction the space $CC(\mathcal{E})$ of universal compatible connections is a real affine space with associated vector space the space of skew adjoint elements of $\text{Hom}_{\mathcal{A}}(\mathcal{E}, \mathcal{E} \otimes_{\mathcal{A}} \Omega^1)$. Here $\Omega^1$ is endowed with the involution $(adb)^* = (db^*)a^*$, $\forall a, b \in \mathcal{A}$. For $\mathcal{E} = \mathcal{A}$ a universal compatible connection is given by an element $\rho$ of $\Omega^1$ such that $\rho^* = -\rho$.

Let then $\Omega^*(\mathcal{A})$ be the universal differential graded algebra of $\mathcal{A}$, reduced by the condition $d1 = 0$ (cf. [8]). It is the quotient of the universal differential graded algebra used in [11] by the two-sided ideal generated by $d1$. Every element of $\Omega^n(\mathcal{A})$ is a linear combination of elements of the form $a^0 da^1 \ldots da^n$ and one has $d(a^0 da^1 \ldots da^n) = da^0 da^1 \ldots da^n = 1 da^0 \ldots da^n$. We extend the above involution of $\Omega^1(\mathcal{A})$ to $\Omega^*(\mathcal{A})$ so that $(a^0 da^1 \ldots da^n)^* = d(a^n)^* d(a^{n-1})^* \ldots d(a^1)^* a^{0*}$. In particular for $\rho \in \Omega^1(\mathcal{A})$ one gets:

$$d(\rho^*) = -(d\rho)^*.$$

For $\mathcal{E} = \mathcal{A}$ and a universal compatible connection $\rho = -\rho^* \in \Omega^1$ its curvature is the selfadjoint element of $\Omega^2$ given by $\theta = d\rho + \rho^2$. In general, as in [8], the curvature $\theta$ of a universal compatible connection $\nabla$ on $\mathcal{E}$ is the endomorphism $\nabla^2$ of the induced module: $\tilde{\mathcal{E}} = \mathcal{E} \otimes_{\mathcal{A}} \Omega^*(\mathcal{A})$.

One has a natural action (cf. [7]) of unitary endomorphisms of $\mathcal{E} : \mathcal{U} = \{u \in \text{End}_{\mathcal{A}}(\mathcal{E}); u^*u = uu^* = 1\}$ on the affine space of universal compatible connections. It is given by

$$\gamma_u(\nabla)\xi = u\nabla(u^*\xi) \quad \forall \xi \in \mathcal{E}.$$

For $\mathcal{E} = \mathcal{A}$ it gives $\gamma_u(\rho) = ud(u^*) + u\rho u^*$, $\forall \rho \in \Omega^1$.

Finally the curvature of $\gamma_u(\nabla)$ is equal to $u\theta u^*$ where $\theta$ is the curvature of $\nabla$.

**Lemma 2** *Let $(\mathfrak{H}, D, \gamma)$ be a $4^+$ summable K-cycle over $\mathcal{A}$, then the following equality defines a positive trace $\tau$ on $\Omega^{\text{ev}}(\mathcal{A})$:*

$$\tau(a^0 da^1 \ldots da^4) = \text{Tr}_w((1+\gamma)a^0[D, a^1][D, a^2][D, a^3][D, a^4]D^{-4}).$$

By construction $\tau$ depends only on the positive cocycle $\phi_w$ given by the equality $(*)$ above.

One defines a $*$ homomorphism $\pi$ of the involutive algebra $\Omega(\mathcal{A})$ in the algebra $\mathcal{L}(\mathfrak{H})$ of bounded operators in $\mathfrak{H}$ by

$$\pi(a^0 da^1 \ldots da^k) = a^0 i[D, a^1] i[D, a^2] \ldots i[D, a^k] \quad \forall a^j \in \mathcal{A}.$$

The image of $\pi$ is contained in the centralizer of the positive linear functional $\psi$ on $\mathcal{L}(\mathfrak{H})$ given by:

$$\psi(T) = \text{Tr}_w(TD^{-4}) \quad \forall T \in \mathcal{L}(\mathfrak{H}).$$

Since $1+\gamma$ commutes with $D^{-4}$, it is also in the centralizer of $\psi$; moreover it commutes with $\pi(\rho)$ for any $\rho \in \Omega^{\text{ev}}$. It follows that $\tau(\rho) = \psi((1+\gamma)\pi(\rho))$ is a positive trace on $\Omega^{\text{ev}}(\mathcal{A})$.

**Proposition 1** *Let $\mathcal{E}$ be a hermitian finite projective module over $\mathcal{A}$. The following equality defines a positive quartic and gauge invariant action functional on the space $CC(\mathcal{E})$ of universal compatible connections:*

$$I(\nabla) = \tau(\theta^2).$$

In this expression we have extended the trace $\tau$ on $\Omega^{\text{ev}}(\mathcal{A})$ to a trace on the endomorphisms of the induced module: $\mathcal{E} \otimes_\mathcal{A} \Omega^{\text{ev}}(\mathcal{A})$. By construction this action functional depends only on the positive cocycle $\phi_w$.

When $\mathcal{E} = \mathcal{A}$ the action $I$ is given by:

$$I(\rho) = \tau((d\rho + \rho^2)^2) \quad \forall \rho = -\rho^* \in \Omega^1(\mathcal{A}).$$

The value of $I(\rho)$ depends only on the pair of bounded operators $\pi(\rho)$ and $\pi(d\rho)$ in $\mathfrak{H}$ both of which belong to the centralizer of $\psi$. Since $\psi$ defines a *finite positive trace* on its centralizer we can use the classical $L^p$ norms ([13]) for both $\pi(\rho)$ ad $\pi(d\rho)$ to estimate $I(\rho)$. In particular:

$$\begin{aligned}(\|\pi(\rho)\|_2)^2 &= \text{Tr}_w(\pi(\rho^*\rho)D^{-4}), \\ (\|\pi(d\rho)\|_2)^2 &= \text{Tr}_w(\pi((d\rho)^*d\rho)D^{-4}).\end{aligned}$$

If both of them vanish then $I(\rho) = 0$. It is important to stress that in general $\pi(d\rho)$ cannot be computed from $\pi(\rho)$. This is already apparent when $\mathcal{A} = C^\infty(M)$ is the algebra of smooth functions on a compact 4-dimensional Riemannian manifold and $(\mathfrak{H}, D, \gamma)$ the $K$-cycle given by the Dirac operator. We shall see that in this case we recover the usual Yang-Mills action on usual connections but the term $d\rho$ introduces an additional *Gaussian field* which can be eliminated. Let us restrict for that discussion to the trivial bundle, so that $\mathcal{E} = \mathcal{A} = C^\infty(M)$. The bimodule $\Omega^1(\mathcal{A})$ consists of functions $\rho(x,y)$ on $M \times M$ which vanish on the diagonal; to $\sum a\,db$ corresponds $\rho(x,y) = \sum a(x)(b(y) - b(x))$.

For $\rho = \sum a\,db \in \Omega^1(\mathcal{A})$, its image under the homomorphism $\pi$ is the operator in $\mathfrak{H} = L^2(M,S)$ given by the Clifford multiplication by the usual differential form $A = \sum a\,d_M b$, where $d_M$ is the usual de Rham differential. In particular one has:

$$\|\pi(\rho)\|_2 = \|A\|_2,$$

where the right hand side is the usual $L^2$ norm of the vector potential $A$ using the Riemannian metric and the volume form. The same equality holds for any $p$ with the corresponding $L^p$ norms on both sides.

However $\pi(d\rho)$ contains more information than the differential 2-form $d_M(A)$. Indeed for $\rho = \sum a\,db$ one sees that $\pi(d\rho) = \sum \pi(da)\pi(db) = \sum \gamma(d_M a)\gamma(d_M b)$ is the multiplication by the section $\sum d_M a.d_M b$ of the Clifford algebra bundle. The orthogonal projection of this section on the scalar sections is $\sum \langle d_M a, d_M b\rangle = \alpha$, which is an additional scalar field. One thus gets:

$$\pi(d\rho) = \alpha + \gamma(dA),$$

$$\|\pi(d\rho)\|_2^2 = \|\alpha\|_2^2 + \|dA\|_2^2.$$

The action $I(\rho)$ then splits into two parts:

$$I(\rho) = \|d_M A + A \wedge A\|_2^2 + \|\langle A, A\rangle + \alpha\|_2^2,$$

where the first part is the usual Yang-Mills action[3] and the second part is Gaussian in $\alpha$ with minimum at $\alpha = -\langle A, A\rangle$. It is thus equivalent to the Yang-Mills action.

To be able to write down the Fermionic Lagrangian in full generality we are just missing the term $\bar\psi \slashed{\partial}_A \psi$ involving spinor fields $\psi$ and the Dirac operator $\slashed{\partial}_A$ with the vector potential $A$. This will be achieved by the following lemma:

**Lemma 3** *Let $\mathcal{A}, \mathcal{E}, (\mathfrak{H}, D, \gamma)$ be as above.*
*1) The tensor product $\mathcal{E} \otimes_{\mathcal{A}} \mathfrak{H}$ is a Hilbert space with inner product $\langle \xi_1 \otimes \eta_1, \xi_2 \otimes \eta_2\rangle = \langle \eta_1, \langle \xi_1, \xi_2\rangle \eta_2\rangle$.*
*2) For any universal compatible connection $\nabla \in CC(\mathcal{E})$ the following equality defines a selfadjoint operator in the Hilbert space $\mathcal{E} \otimes_{\mathcal{A}} \mathfrak{H}$:*

$$D_\nabla(\xi \otimes \eta) = \xi \otimes D\eta - i((1 \otimes \pi)\nabla \xi)\eta, \quad \forall \xi \in \mathcal{E}, \ \eta \in \mathfrak{H}.$$

When $\mathcal{E} = \mathcal{A}$ and $\nabla$ is associated to $\rho = -\rho^* \in \Omega^1(\mathcal{A})$ the above operator is just $D - i\pi(\rho)$. For $\rho = \sum a\,db$ this gives $D + \sum a[D, b]$.

One also checks that the operator associated to $\gamma_u(\nabla)$ is given by $uD_\nabla u^*$, for any $u \in \mathcal{U}_\mathcal{E} = \{u \in \text{End}_\mathcal{A}\mathcal{E};\ uu^* = u^*u = 1\}$.

**Theorem 1** *For any value of the coupling constants $\lambda$, $m$ the following expression is a gauge invariant (i.e. invariant under the group $\mathcal{U}_\mathcal{E}$) action on pairs $\nabla \in CC(\mathcal{E})$ and $\psi \in \mathcal{E} \otimes_\mathcal{A} \mathfrak{H}$:*

$$\mathcal{L}(\nabla, \psi) = I(\nabla) + \lambda \langle \psi, D_\nabla \psi\rangle + m\langle \psi, \psi\rangle.$$

---
[3] Multiplied by $(8\pi^2)^{-1}$, and where we kept the $A \wedge A$ term to indicate what would occur for higher dimensional bundles.

## 2. Essay on Physics and Non-commutative Geometry

This expression with suitable choice of $\mathcal{E}$ gives back the usual Lagrangian in Euclidean formulation in the special case where $\mathcal{A}$ is the algebra of functions on Euclidean space and $(\mathfrak{H}, D, \gamma)$ the Dirac $K$-cycle. It ought to be the point of departure of an analysis of what physics could be when associated with a non-commutative space. The problems of choosing $\mathcal{A}$ and then of choosing $D$ once $\mathcal{A}$ is fixed will be discussed in the next section.

In order to make sure that the action $I(\nabla)$ is not trivial we shall prove a general inequality similar to the inequality $c_2(E) + c_1(E)^2 \leq$ YM($\nabla$) between the Chern classes of a hermitian vector bundle on a compact 4-manifold and the value of the Yang-Mills action on connections.

We recall from Theorem 4 of Section 3 that the Hochschild class of the character of a $4^+$ summable $K$-cycle is given by the Hochschild cocycle:

$$(**) \quad \psi_w(a^0, \ldots, a^4) = \operatorname{Tr}_w(\gamma a^0 [D, a^2] \ldots [D, a^4] D^{-4}).$$

We know that $B\psi_w = 0$ in the group $HC^3(\mathcal{A})$ of cyclic 3-cocycles on $\mathcal{A}$, as follows from the long exact sequence of Theorem 3 of Section 3. We shall make the further hypothesis that $B\psi_w$ is not only a coboundary but is actually equal to 0. It follows then that $\psi_w$ has a well defined class:

$$[\psi_w] \in HC^4(\mathcal{A}) = \operatorname{Ker} b \cap \operatorname{Ker} B / b(\operatorname{Ker} B). \quad \text{(cf. [8])}$$

Our inequality will involve this class, which differs in general from the cyclic cohomology class of the $K$-cycle $(\mathfrak{H}, D, \gamma)$.

**Theorem 2** *Let $\mathcal{A}$ be a $*$-algebra, $(\mathfrak{H}, D, \gamma)$ a $4^+$ summable $K$-cycle on $\mathcal{A}$ and $\psi_w$ the Hochschild cocycle $(**)$. Assume that $B\psi_w = 0$. Then for any finite projective module $\mathcal{E}$ over $\mathcal{A}$ one has:*

$$\langle [\mathcal{E}], [\psi_w] \rangle \leq \frac{1}{2} I(\nabla)$$

*for any universal compatible connection $\nabla \in CC(\mathcal{E})$.*

The left hand side is the pairing between cyclic cohomology and $K$ theory. In the case of the Dirac operator on a compact spin manifold $M$, the cocycle $\psi_w$ is $(8\pi^2)^{-1}$ times the fundamental class of $M$ and the action $I(\nabla)$ is $(8\pi^2)^{-1}$ times the sum of the usual Yang-Mills action with the topological action $\int_M \theta^2$, so that the above inequality is the usual one: $c_1(E)^2 + c_2(E) \leq$ YM($\nabla$) for any compatible connection $\nabla$ on the vector bundle $E$.

## 5. Discussion of Some Models

What we have done so far is to recover (Theorem 2 of Section 4) the Euclidean action functional $\mathcal{L}(A, \psi)$ from the following data:

1) the $*$ algebra $\mathcal{A}$ of functions on Euclidean space-time,

2) the Dirac operator as a $K$-cycle on $\mathcal{A}$.

Moreover we did it in such a way that $\mathcal{A}$ was just required to be a $*$ algebra, not necessarily commutative, and the $K$-cycle $(\mathfrak{H}, D, \gamma)$ was just required to be $4^+$ summable.

This just shows that provided we understand how to determine the $K$-cycle $(\mathfrak{H}, D, \gamma)$ given $\mathcal{A}$, the commutativity of coordinates in Euclidean space-time or the manifold structure of this space is not a mandatory requirement in order to formulate a sensible point of departure for field theory. Two fundamental questions remain unanswered:

a) Under which conditions will the resulting field theory be renormalisable?

b) Why should one gain anything by dealing with more elaborate spaces?

Our strategy will be to ignore problem a) for the time and worry about problem b) but working at the classical level. We shall consider what classical estimates we get for the masses such as the Higgs mass, bearing in mind that the quantum meaning is unclear.

In order to test our ideas we shall concentrate on the well established part of particle physics, namely the electroweak interaction unified in the Glashow–Weinberg–Salam model.

Our aim is thus to modify the usual picture of Euclidean space-time and get a new picture (cf. Fig. 2) which is within our framework and whose action functional and field content[4] is that of the Weinberg–Salam model. This is just a starting point so that we concentrate on a conceptual understanding of the Higgs fields, the "black box" of the standard model.

The qualitative picture that emerges from our considerations is that of a two-sheeted Euclidean space-time, whose two sheets $M$ and $M'$ are extremely close, every point $x$ of $M$ being at a distance of $\sim 10^{-16}$ cm from some point $x'$ of $M'$. Thus we are dealing with the simplest possible Kaluza–Klein theory where the fibre is: two points! Of course ordinary differential geometry does not do anything with

---

[4] As prescribed in Theorem 2 of Section 4.

## 2. Essay on Physics and Non-commutative Geometry

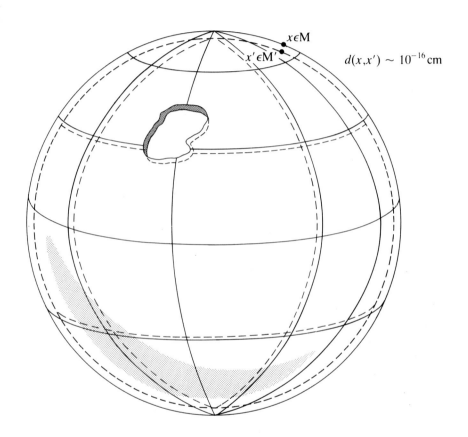

Fig. 2: A double-sheeted Euclidean space-time.

a two-point space but non-commutative differential geometry does. (Even though the algebra $\mathcal{A}$ we shall be dealing with is *commutative* the abandon of local charts and replacement by operator theoretic data gives much more freedom to manoeuvre.) Thus the Higgs fields will appear essentially from the quantized differential $\frac{f(x')-f(x)}{\ell}$ of a function on $X = M \cup M'$, where $\ell$ is the distance between the two sheets, and the disconnectedness of the fibre will be used to get non-trivial bundles on $X$ *whose dimensions differ* on the two copies of $M$.

It is clear that the models that we put forward in this section, namely models II and III, need a lot of improvement before they can have a real physical content, but at least they are very close to the Weinberg–Salam model except for constraints on the coupling constants and masses including a Higgs mass of roughly twice the mass of the $W$, for model II.

Model I is obviously wrong but it serves as a warm up exercise for the understanding of the other two.

These models will only use *commutative* $*$ algebras $\mathcal{A}$. Our first point is that a $K$-cycle $(\mathfrak{H}, D, \gamma)$ over $\mathcal{A}$ determines a *metric d* on the spectrum $X$ of $\mathcal{A}$ (i.e. elements of $\mathcal{A}$ are complex-valued functions on $X$) by the formula:

$$d(p,q) = \sup\{|f(p) - f(q)| : f \in \mathcal{A}, \|[D,f]\| \leq 1\}.$$

We have seen above that this distance is the Riemannian geodesic distance on a manifold $M$ when $D = \emptyset_M$ is the Dirac operator. There are however many interesting metric spaces which can be obtained this way and our models will be found by first guessing a suitable modification of Euclidean space-time as a metric space. In order to ignore at first the infrared divergences we stick to compact spaces.

Model I is obviously not very interesting but it already exhibits the Higgs field and minimal interaction.

*Model I*

Let $M$ be a 4-dimensional compact spin Riemannian manifold and $L$ some given length of the order of the diameter of $M$. Consider the *metric space* $(X,d)$, where the set $X$ is the union of $M$ and an additional point $v \notin M$. The distance $d$ is uniquely defined by the following conditions:

$$\begin{aligned} d(v,p) &= L, \quad \forall p \in M, \\ d(p,q) &= \inf(d_M(p,q), 2L), \quad \text{for } p,q \in M. \end{aligned}$$

(Here we used $d_M$ for the geodesic distance on $M$.) This space $(X,d)$ is of course compact and we let $\mathcal{A}$ be the $*$ algebra of functions on $X$.

## 2. Essay on Physics and Non-commutative Geometry

We write every $a \in \mathcal{A}$ as a pair $(f, \lambda)$, where $f$ is the restriction of $a$ to $M \subset X$ and $\lambda$ is the scalar $a(v)$.

We now describe a $K$-cycle on $\mathcal{A}$ with associated metric equal to $d$ (and which was defined in [8, Part I]). We let $\mathcal{A}$ act in the Hilbert space $\mathfrak{H} = L^2(M, S) \otimes \mathbb{C}^2$ by $a \mapsto \begin{bmatrix} f & 0 \\ 0 & \lambda \end{bmatrix}$. We then take $D = \partial\!\!\!/_M \otimes \sigma_3 + \mu \otimes \sigma_1$, $\gamma = \gamma_M \otimes \sigma_3$, where $\partial\!\!\!/_M$ is the Dirac operator on $M$, $\mu = \frac{1}{L}$ and the $\sigma_j$'s are the Pauli $2 \times 2$ matrices. One checks:

**Lemma 1** $(\mathfrak{H}, D, \gamma)$ *is a $4^+$ summable $K$-cycle on $\mathcal{A}$ whose associated metric on $X$ is equal to $d$.*

The $4^+$ summability follows from the equality $D^2 = (\partial\!\!\!/_M{}^2 + \mu^2) \otimes 1$.

We can thus apply the general formulae of Section 4 and determine precisely the vector potentials and the action. We take the trivial finite projective module $\mathcal{E} = \mathcal{A}$. Let thus $\rho = -\rho^* \in \Omega^1(\mathcal{A})$. With the notations of Section 4 we need to determine the operators:

$$\pi(\rho) = \sum a_j i[D, b_j] \quad \text{and} \quad \pi(d\rho) = \sum i[D, a_j] \, i[D, b_j],$$

as well as the action:

$$I(\rho) = \operatorname{Tr}_w((1 + \gamma)\theta^2 D^{-4}), \quad \theta = \pi(d\rho) + \pi(\rho)^2.$$

We shall identify sections of the Clifford algebra bundle on $M$ with the corresponding operators in $L^2(M, S)$. One then gets the following equalities:

$$\pi(\rho) = \begin{bmatrix} A & i\mu\phi \\ i\mu\bar\phi & 0 \end{bmatrix}, \pi(d\rho) = \begin{bmatrix} dA + \psi - \mu^2(\phi + \bar\phi) & i\mu(A + d\phi) \\ i\mu(A - d\bar\phi) & -\mu^2(\phi + \bar\phi) \end{bmatrix},$$

where the fields $A, \phi, \psi$ are respectively an ordinary vector potential on $M$ (i.e. a 1-form $A$ such that $A^* = -A$), a complex scalar field on $M$ and a real scalar field. Also $dA$ is the ordinary 2-form differential of $A$ viewed as a section of the Clifford algebra bundle by the inclusion $\Lambda^2 T_x^* \subset \operatorname{Cliff}(T_x^*)$. In terms of $a_j = (f_j, \lambda_j)$, $b_j = (g_j, \mu_j)$ the fields $A, \phi, \psi$ are:

1) $A = \sum f_j dg_j \in A^1(M)$,

2) $\phi = -\sum f_j(g_j - \mu_j)$,

3) $\psi = \frac{1}{2} \sum \langle df_j, dg_j \rangle$.

The freedom in the choice of $\rho$ shows that $A$, $\phi$, $\psi$ vary independently, i.e. we obtain any triple $A^* = -A$, $\psi = \psi^*$.

Using [6] we see that the value of the Dixmier trace

$$I(\rho) = \mathrm{Tr}_w((1+\gamma)\theta^2 D^{-4})$$

is independent of $w$, and up to the topological term $\mathrm{Tr}_w(\gamma\theta^2 D^{-4})$ which is identically 0, $I(\rho)$ is just the integral over the manifold $M$ of the square of the Hilbert–Schmidt norm of the following $2 \times 2$ matrix of elements of the Clifford algebra:

$$\theta = \pi(d\rho) + \pi(\rho)^2 = \begin{bmatrix} dA + A^2 + \psi - \mu^2(\phi\bar\phi + \phi + \bar\phi) & i\mu(A + A\phi + d\phi) \\ i\mu(A + A\bar\phi - d\bar\phi) & -\mu^2(\phi\bar\phi + \phi + \bar\phi) \end{bmatrix}.$$

The real scalar field $\psi$ appears as a Gaussian in $I(\rho)$ and we eliminate it by taking the minimum. We then get:

**Proposition 1** *Once the Gaussian field $\psi$ is eliminated the action $I(\rho)$ is given by the following formula:*

$$I(\rho) = (8\pi^2)^{-1} \int \mathcal{L}(A,\phi);$$

$$\mathcal{L}(A,\phi) = |dA + A \wedge A|^2 + 2\mu^2 |(\phi+1)A + d\phi|^2 + \mu^4(\phi\bar\phi + \phi + \bar\phi)^2.$$

(We kept $A \wedge A$ to indicate what would happen if instead of $\mathcal{E} = \mathcal{A}$ we had taken $\mathcal{E} = \mathcal{A}^k$ for some $k$.)

Thus if we change variables to $\phi+1$ we get the Yang–Mills–Higgs Lagrangian; the second term is the minimal coupling $|d_A(\phi+1)|^2$ and the last term is $||\phi+1|^2 - 1|^2 = (\phi\bar\phi + \phi + \bar\phi)^2$.

This is a satisfactory indication but this model has a fatal drawback: *it does not have the correct Fermionic sector*. If we consider the action of $\mathcal{A}$ in $\mathfrak{H}$ it corresponds to a bundle on $X$ whose restriction to $M$ is the spinor bundle but whose fibre at $v \in X$ is infinite dimensional, equal to $L^2(M,S)$. This implies that gauge transformations do not act on this copy of $L^2(M,S)$ which is unacceptable.

*Model II*

If in the metric space $X = M \cup \{v\}$ of the first model, we want the restriction of the metric $d$ to $M$ to be the Riemannian metric $d_M$, then we need to take $L$ large enough: $L \geq \frac{1}{2} \mathrm{diameter}(M)$. This lower bound is tied up with M. Gromov's notion of distance on the set of metric spaces [10]. Given two metric spaces $(M, d_M)$ and $(N, d_N)$ their Gromov distance is the infimum of the Hausdorff distance of $M$

## 2. Essay on Physics and Non-commutative Geometry

and $N$ as subsets of the metric space $X = M \cup N$, whose metric $d_X$ restricts to $d_M$ on $M$ and to $d_N$ on $N$. The above lower bound on $L$ expresses the trivial fact that $\text{dist}(M, \text{pt.}) = \frac{1}{2} \text{diameter}(M)$.

Since a lower bound on $L$ gives an upper bound on $\mu = \frac{1}{L}$, this limits considerably the interest of Model I. On the other hand, for the Gromov metric one has $\text{dist}(M, M) = 0$, and the construction of Model II will be based on the corresponding metric space. Thus we let $\ell$ be a given small length and we endow $X = M \cup M$, the union of two copies of $M$, with the following metric (see Fig. 2):

$$\begin{aligned} d(x, y) &= d_M(x, y) & \text{for } x, y \in M, \\ d(x', y') &= d_M(x, y) & \text{for } x', y' \in M' \text{ (the other copy of } M\text{)}, \\ d(x, y') &= d_M(x, y) + \ell & \text{for } x \in M, \ y' \in M', \end{aligned}$$

where we used the notation $x$ to indicate a point of $M$ and $x'$ for its image in the second copy $M'$ of $M$.

We let $\mathcal{A}$ be the $*$ algebra of functions on $X$ and write every $a \in \mathcal{A}$ as a pair $a = (f, f')$ of functions on $M$.

We now describe a $K$-cycle on $\mathcal{A}$ whose associated metric is equal to $d$. The triple $(\mathfrak{H}, D, \gamma)$ is the same as in Model I. The only difference comes from the action of $\mathcal{A}$, given by:

$$a \mapsto \begin{bmatrix} f & 0 \\ 0 & f' \end{bmatrix} \quad \text{acting in } \mathfrak{H} = L^2(M, S) \otimes \mathbb{C}^2.$$

**Lemma 2** *The triple $(\mathfrak{H}, D, \gamma)$ is a $4^+$ summable $K$-cycle on $\mathcal{A}$ whose associated metric is equal to $d$.*

This is easy to check since the operator $i[D, a]$ is given by the following $2 \times 2$ matrix of sections of the Clifford algebra bundle:

$$i[D, a] = \begin{bmatrix} df & i\mu(f' - f) \\ i\mu(f - f') & df' \end{bmatrix}, \quad \mu = \frac{1}{\ell},$$

so that $\|[D, a]\| \leq 1$ is equivalent to:

$$|f(x) - f(y)| \leq d_M(x, y), \quad |f'(x) - f'(y)| \leq d_M(x, y), \quad |f(x) - f'(x)| \leq \ell,$$

for any $x, y \in M$.

We are now going to compute the action $\text{Tr}_w((1 + \gamma)\theta^2 D^{-4})$, but this time we shall use a *non-trivial* bundle $E$ on $X$. This bundle is trivial with 1-dimensional fibre when restricted to the first copy of $M$, and trivial with 2-dimensional fibre when restricted to the second copy $M'$ of $M$ in $X$. It is a subbundle of the trivial bundle with fibre $\mathbb{C}^2$ everywhere, and one has:

$$\mathcal{E} = e\mathcal{A}^2, \quad e = \begin{bmatrix} (1,1) & 0 \\ 0 & (0,1) \end{bmatrix} \in M_2(\mathcal{A}).$$

The non-triviality of this bundle comes from the presence of the finite difference terms $i\mu(f - f')(x) = i\frac{f(x)-f'(x)}{\ell}$ which occur in the off-diagonal terms of $i[D,a]$. It is thus clear that the idempotent $e$ defining $\mathcal{E}$ does not yield a flat connection on $\mathcal{E}$. Any universal compatible connection $\nabla$ on $\mathcal{E}$ (cf. Section 4) is of the form:

$$\nabla \xi = ed\xi + \rho\xi,$$

where $\rho = -\rho^*$ is an element of $M_2(\Omega^1(\mathcal{A}))$ such that $e\rho = \rho e$. If we write $\rho = \begin{pmatrix} \rho_{11} & \rho_{12} \\ \rho_{21} & \rho_{22} \end{pmatrix}$, this condition means that $(0,1)\rho_{21} = \rho_{21}$, $(0,1)\rho_{22} = \rho_{22} = \rho_{22}(0,1)$.

The curvature $\theta = \nabla^2$ is given by $\theta = e\,de\,de + e\,d\rho + \rho^2$.

Our task is to compute $\pi(\rho)$, $\pi(e\,d\rho)$ and $\mathrm{Tr}_w((1+\gamma)\theta^2 D^{-4})$. Each $\pi(\rho_{kl})$ is a $2\times 2$ matrix so that $\pi(\rho)$ is a $4\times 4$ matrix, but since $e\rho = \rho e$ with $e = \begin{bmatrix} 1 & 0 & 0 & 0 \\ 0 & 1 & 0 & 0 \\ 0 & 0 & 0 & 0 \\ 0 & 0 & 0 & 1 \end{bmatrix}$ the third line and column are zero and both $\pi(\rho)$ and $\pi(e\,d\rho)$ are just $3 \times 3$ matrices. A tedious computation gives the following formulae:

$$\pi(\rho) = \begin{bmatrix} A_1 & i\mu\bar{\phi}_1 & i\mu\bar{\phi}_2 \\ i\mu\phi_1 & -A'_1 & W^* \\ i\mu\phi_2 & -W & -A_2 \end{bmatrix},$$

where $A_1$, $A'_1$, $A_2$ are ordinary vector potentials (skew-adjoint 1-forms on $M$); $W$ is a complex 1-form, and $\phi_1$, $\phi_2$ are two complex scalar fields. With $\rho_{kl} = \sum a_{klj}\,db_{klj}$, $a_{klj} = (f_{klj}, f'_{klj})$, $b_{klj} = (g_{klj}, g'_{klj})$, one has:

$$\begin{aligned}
A_1 &= \sum f_{11j}\,dg_{11j}, & A'_1 &= \sum f'_{11j}\,dg'_{11j}, \\
A_2 &= \sum f'_{22j}\,dg'_{22j}, & W &= \sum f'_{21j}\,dg'_{21j}, \\
\bar{\phi}_1 &= \sum f_{11j}(g'_{11j} - g_{11j}), & \phi_2 &= \sum f'_{21j}(g_{21j} - g'_{21j}).
\end{aligned}$$

We then get the following formula for $\pi(e\,d\rho)$:

$$\pi(e\,d\rho) = \begin{bmatrix} dA_1 + \psi_1 - \mu^2(\phi_1 + \bar{\phi}_1) & & \\ i\mu(A_1 - A'_1 - d\phi_1) & dA'_1 + \psi'_1 - \mu^2(\phi_1 + \bar{\phi}_1) & \\ -i\mu(W + d\phi_2) & -\mu^2\phi_2 + dW + \psi & dA_2 + \psi_2 \end{bmatrix},$$

## 2. Essay on Physics and Non-commutative Geometry

where using the self-adjointness of $\pi(e\,d\rho)$ we did not write the upper right part of the matrix. Here the fields $\psi_1$, $\psi_1'$, $\psi_2$ are real scalar fields and $\psi$ is a complex scalar field, given by:

$$\psi_1 = \tfrac{1}{2}\sum \langle df_{11j}, dg_{11j}\rangle, \qquad \psi_1' = \tfrac{1}{2}\sum \langle df'_{11j}, dg'_{11j}\rangle$$
$$\psi_2 = \tfrac{1}{2}\sum \langle df'_{22j}, dg'_{22j}\rangle, \qquad \psi = \tfrac{1}{2}\sum \langle df'_{21j}, dg'_{21j}\rangle.$$

All these fields are independent. We then have to compute the various elements of $\theta = \pi(e\,de\,de) + \pi(e\,d\rho) + \pi(\rho)^2$, which is a $3\times 3$ matrix of sections of the Clifford algebra bundle.

$$\theta_{11} = dA_1 + A_1^2 + \psi_1 - \mu^2(\phi_1\bar\phi_1 + \phi_1 + \bar\phi_1 + \phi_2\bar\phi_2),$$
$$\theta_{12} = i\mu((A_1 - A_1')(\bar\phi_1 + 1) - \bar\phi_2 W + d\bar\phi_1), \quad \theta_{21} = \theta_{12}^*,$$
$$\theta_{13} = i\mu(W^*(\bar\phi_1 + 1) + (A_1 - A_2)\bar\phi_2 + d\bar\phi_2), \quad \theta_{31} = \theta_{13}^*,$$
$$\theta_{22} = -\mu^2(\phi_1\bar\phi_1 + \phi_1 + \bar\phi_1) + dA_1' + A_1'^2 - W^*W + \psi_1',$$
$$\theta_{23} = -\mu^2\bar\phi_2(\phi_1 + 1) - dW^* - A_1'W^* - W^*A_2 + \bar\psi, \quad \theta_{32} = \bar\theta_{23},$$
$$\theta_{33} = dA_2 + A_2^2 - WW^* + \psi_2 + \mu^2(1 - \phi_2\bar\phi_2).$$

The value of $I(\rho) = \operatorname{Tr}_w((1+\gamma)\theta^2 D^{-4})$ is the integral over $M$ of the sum of the square norms of the $\theta_{ij}$. We thus see several very interesting terms.

First if we look only at the components of degree 2 in the Clifford algebra, the matrix $\theta$ reduces to:

$$\begin{bmatrix} dA_1 & 0 & 0 \\ 0 & dA_1' - W^*\wedge W & -dW^* - (A_1' - A_2)\wedge W^* \\ 0 & dW + W\wedge(A_1' - A_2) & dA_2 - W\wedge W^* \end{bmatrix}.$$

Thus their contribution to the Lagrangian is just the sum of square norms of the curvatures of the ordinary compatible connections $A_1$ and $\begin{bmatrix} A_1' & -W^* \\ W & A_2 \end{bmatrix} = T.$

Next we look only at the components of degree 1, i.e. the odd part, in the Clifford algebra; the matrix $\theta$ then reduces to

$$\begin{bmatrix} 0 & \theta_{12} & \theta_{13} \\ \theta_{21} & 0 & 0 \\ \theta_{31} & 0 & 0 \end{bmatrix},$$

and their contribution is $2\mu^2(d_X\phi)^2$, where $\phi$ is the doublet $(\phi_1+1, \phi_2)$ and $d_X\phi$ is the covariant derivative: $d_X\phi = d\phi + X\phi$, where $X = T - A_1 1$ is the $2\times 2$ matrix

$$X = \begin{bmatrix} A_1' - A_1 & -W^* \\ W & A_2 - A_1 \end{bmatrix}.$$

We should now carefully compare these two terms with the boson sector of the Weinberg–Salam model (cf. [4] p. 287). Since instead of a $U(2) \times U(1)$ theory this model is an $SU(2) \times U(1)$ theory, we impose the 0-trace condition:

$$(*) \quad A_1 = A_1' + A_2.$$

We then make the following changes of variables to match the notations of [4]:

$$(**) \quad iA_1 = g_1 B, \quad iA_1' = \frac{g}{2} A^3 + \frac{g_1}{2} B, \quad iA_2 = -\frac{g}{2} A^3 + \frac{g_1}{2} B,$$

$$W = g(A^1 - iA^2), \quad \phi = \frac{g}{\sqrt{2}\mu} \tilde{\phi}.$$

Then the above contribution to the lagrangian is given by[5]:

$$g_1^2(1 + \frac{1}{2})|dB|^2 + \frac{1}{2}g^2|G|^2 + g^2|\partial_{(A^a,B)}\tilde{\phi}|^2,$$

where $dB$ is the field strength tensor of the abelian field $B$ and $G$ the field strength tensor of the non-abelian field $A^a_\nu$. The second factor of $\frac{1}{2}$ in front of $|dB|^2$ comes from the contribution $|dA_1' - W^* \wedge W|^2 + |DA_2 + W^* \wedge W|^2$ which contains $2\frac{g_1^2}{4}|dB|^2$. One checks from the contribution of the same term that the coefficients of $|dA^3|^2$ is $2(\frac{g}{2})^2 = \frac{g^2}{2}$.

Thus if we want to obtain a scalar multiple $g^2 \mathcal{L}_B$ of the bosonic part of the Weinberg–Salam Lagrangian we need to impose on the latter the condition: $g_1^2 \times 3 = g^2$. This gives the value $\theta_W = 30°$ for the Weinberg angle which is defined (cf. [4] p. 286) by $\sin \theta_W = \frac{g_1}{\sqrt{g^2 + g_1^2}} = \frac{1}{\sqrt{4}} = \frac{1}{2}$. This value is in good agreement with the experimental value (cf. [4] p. 292).

The term $\partial_{(A^a,B)}\tilde{\phi}$ is the covariant differential given by

$$(\partial_\mu - \frac{ig}{2}\tau \cdot A_\mu + \frac{ig_1}{2} B_\mu)\tilde{\phi},$$

in terms of the Pauli spin matrices $\tau_j$ (cf. [4] p. 307 for notations). We have corrected the wrong sign in [4] p. 287 formula (5) and used a + sign in front of $\frac{ig_1}{2} B_\mu$.

For this to agree with the corresponding term of $\mathcal{L}_B$ we took $\tilde{\phi} = \sqrt{2}\frac{\mu}{g}\phi$ which gives the following relation between our parameter $\mu = \frac{1}{\ell}$ and the parameter $\eta$ which appears in the last term of $\mathcal{L}_B$ : $\frac{\lambda^2}{4}(|\tilde{\phi}|^2 - \eta^2)$, we get:

$$\eta = \sqrt{2}\frac{\mu}{g},$$

---

[5] Up to an overall factor of $4 = \text{tr}(1)$ in the Clifford algebra.

## 2. Essay on Physics and Non-commutative Geometry

i.e. $\mu = M_W$ the mass of the intermediate boson. Thus we get for the distance $\ell$ an order of magnitude of $10^{-16}$ cm.

All this is fine but we need to compute the last contribution to our action $I(\rho)$, i.e. that of the terms of degree 0 in the Clifford algebra. The matrix $\theta$ then reduces to :

$$\begin{bmatrix} A_1^2 + \psi_1 - \mu^2(\phi_1\bar{\phi}_1 + \phi_1 + \bar{\phi}_1 + \phi_2\bar{\phi}_2) & 0 & 0 \\ 0 & (A_1'^2 - W^*W)_0 + \psi_1' - \mu^2(\phi_1\bar{\phi}_1 + \phi_1 + \bar{\phi}_1) & 0 \\ 0 & (W^*(A_1' - A_2))_0 + \bar{\psi} - \mu^2\bar{\phi}_2(1 + \phi_1) & (A_2^2 - WW^*)_0 + \psi_2 + \mu^2(1 - \phi_2\bar{\phi}_2) \end{bmatrix}.$$

Here we used the notation $(\;)_0$ to indicate the component of degree 0: thus for instance $(A_2^2 - W^*W)_0 = A_2^2 - \frac{1}{2}(W^*W + WW^*)$.

If we compute the coefficient of $\mu^4$ in the square of the Hilbert–Schmidt norm of this matrix we find[6]:

$$\begin{aligned}(|\phi_1 + 1|^2 + |\phi_2|^2 - 1)^2 &+ (|\phi_1 + 1|^2 - 1)^2 \\ &+ 2|\phi_2|^2|\phi_1 + 1|^2 + (1 - |\phi_2|^2)^2 \\ &= 2(|\phi_1 + 1|^2 + |\phi_2|^2 - 1)^2 + 1 \\ &= 2(|\phi|^2 - 1)^2 + 1,\end{aligned}$$

in terms of $\phi = (\phi_1 + 1, \phi_2)$ as above.

Thus if we could only neglect the other terms we would get the complete bosonic sector of the Weinberg–Salam model with the following value for the parameter $\lambda$:

$$2\mu^4 \left(\frac{g}{\sqrt{2}\mu}\right)^4 = g^2 \frac{\lambda^2}{4}, \quad \text{i.e. } \lambda = \sqrt{2}\, g.$$

This gives for the Higgs mass the value $m_\sigma = 2M_W \sim 160\,\text{GeV}$; and it is clear that this value is still compatible with the actual experimental tests. In any case there is no good reason why we could neglect the other terms and if we use the freedom in the additional fields $\psi_1$, $\psi_1'$, $\psi_2$, $\psi$ and take the infimum of the action when they vary we just cancel completely the contribution of this last term. This is a very serious drawback of our Model II but its origin is easy to understand, and will lead us to Model III. The point is that even though we used a non-trivial bundle over $X$ (i.e. the module $\mathcal{E}$ over $\mathcal{A}$) the topological part of our action functional *does not detect the non-triviality of* $\mathcal{E}$. Thus, to be more precise the cyclic cocycle class $[\psi_w]$ where (Section 4):

$$\psi_w(a^0, \ldots, a^4) = \text{Tr}_w(\gamma a^0[D, a^1] \cdots [D, a^4]D^{-4}) \quad \forall a^j \in \mathcal{A},$$

---

[6] Neglecting again the overall factor 4=tr(1).

has a trivial pairing with $\mathcal{E}$: $\psi_w(e,e,e,e,e) = 0$. If this pairing had been $> 0$ then Theorem 2 of Section 4 would have applied to prevent the above cancellation. We shall come back to this point at the beginning of the discussion of Model III.

Let us now compare the fermionic content of Model II with that of the Weinberg–Salam model. With the notation of Section 4 the Hilbert space $\mathcal{E} \otimes_{\mathcal{A}} \mathfrak{H}$ is the direct sum of 3 copies of $L^2(M,S)$ and for a given connection $\nabla = ed + \rho$, the corresponding operator $D_\nabla$ is given by the following $3 \times 3$ matrix of operators in $L^2(M,S)$:

$$D_\nabla = \begin{bmatrix} \partial_M & \mu & 0 \\ \mu & -\partial_M & 0 \\ 0 & 0 & -\partial_M \end{bmatrix} + \frac{1}{i}\pi(\rho),$$

and the $\mathbb{Z}/2$ grading is given by

$$\gamma = \begin{bmatrix} \gamma_M & 0 & 0 \\ 0 & -\gamma_M & 0 \\ 0 & 0 & -\gamma_M \end{bmatrix}.$$

(This follows from the grading $\gamma_M \otimes \sigma_3$ of $\mathfrak{H}$ and the matrix

$$e = \begin{bmatrix} 1 & 0 & 0 & 0 \\ 0 & 1 & 0 & 0 \\ 0 & 0 & 0 & 0 \\ 0 & 0 & 0 & 1 \end{bmatrix}$$

which defines $\mathcal{E} = e\mathcal{A}^2$.)

Thus if we take the right handed part $\frac{1+\gamma}{2}(\mathcal{E} \otimes_{\mathcal{A}} \mathfrak{H})$ of our Hilbert space of fermions we see that it contains a right-handed singlet $R = \frac{1+\gamma_5}{2}\xi$ and a left-handed doublet $L = \frac{1-\gamma_5}{2}\begin{pmatrix} \xi \\ \eta \end{pmatrix}$ exactly as in [4] p. 289. Thus we write the elements of $\mathfrak{H}' = \frac{1+\gamma}{2}(\mathcal{E} \otimes_{\mathcal{A}} \mathfrak{H})$ in the form:

$$\begin{pmatrix} R \\ L \end{pmatrix} = \begin{pmatrix} \xi_+ \\ \xi_- \\ \eta \end{pmatrix}.$$

Recalling that $\frac{1}{i}\pi(\rho) = \begin{bmatrix} -iA_1 & \mu\bar{\phi}_1 & \mu\bar{\phi}_2 \\ \mu\phi_1 & iA'_1 & -iW^* \\ \mu\phi_2 & iW & iA_2 \end{bmatrix}$,

we get:

$$D_\nabla = \begin{bmatrix} \partial_M - iA_1 & \mu(1+\bar{\phi}_1) & \mu\bar{\phi}_2 \\ \mu(1+\phi_1) & -\partial_M + iA'_1 & -iW^* \\ \mu\phi_2 & iW & -\partial_M + iA_2 \end{bmatrix}.$$

## 2. Essay on Physics and Non-commutative Geometry

Thus we get the following contribution from $D_\nabla$ to the Lagrangian, with the notation $\emptyset = \frac{1}{i}\gamma^\mu \partial_\mu$ in the local coordinates on $M$:

$$\begin{aligned}\mathcal{L}(x) &= -i\bar{R}(x)\gamma^\mu(\partial_\mu - ig_1 B_\mu)R(x) \\&+ i\bar{L}(x)\gamma^\mu(\partial_\mu - \frac{ig}{2}\tau \cdot A_\mu - \frac{ig_1}{2}B_\mu)L(x) \\&+ \frac{g}{\sqrt{2}}(\bar{R}(x)(\bar{\phi}^*(x)L(x)) + (\bar{L}(x)\bar{\phi}(x))R(x)).\end{aligned}$$

Thus we get exactly the correct terms of [4] p. 289, with the value $G = \frac{g}{\sqrt{2}}$ for the Yukawa coupling constant. This yields the value $m = \eta \frac{G}{2} = \sqrt{2}\frac{N}{g}\frac{g}{2\sqrt{2}} = \frac{1}{2}M_W \sim 40$ GeV which is much heavier than the heavy $\tau$ lepton $\sim 1.8$ GeV. Thus this model has several bad drawbacks, the main ones are:

1) The absence of a topological term to detect the non-triviality of the module $\mathcal{E}$.

2) A wrong number of leptons with wrong masses.

We shall now describe a third model whose construction is based on the above geometric picture of two copies of Euclidean space-time and the solution of problem 1).

*Model III*

Let us keep the qualitative picture of Model II, so that the space $X = M \cup M$ is the same but the metric $d$ might be different but qualitatively the same (cf. Fig. 2).

The algebra $\mathcal{A}$ and the bundle $E$ (with associated module $\mathcal{E}$) are the same as in Model II.

To solve problem 1) above we need to find a $4^+$ summable $K$-cycle $(\mathfrak{H}, D, \gamma)$ over $\mathcal{A}$ which fulfills the conditions of Theorem 2 of Section 4 above. Thus, with:

$$\psi_w(a^0, a^1, a^2, a^3, a^4) = \text{Tr}_w(\gamma a^0[D, a^1]\cdots[D, a^4]D^{-4}), \quad \forall a^j \in \mathcal{A},$$

we need to have $B\psi_w = 0$ and: $\langle \mathcal{E}, [\psi_w]\rangle > 0$.

Let then $\delta$ be the idempotent, $\delta = \delta^* = \delta^2 \in \mathcal{L}(\mathfrak{H})$, which is given by the action of $(0,1) \in \mathcal{A}$ in the Hilbert space $\mathfrak{H}$. One has $e = 1 \oplus \delta$ and hence: $\langle[\mathcal{E}], \psi_w\rangle = \psi_w(\delta, \delta, \delta, \delta, \delta)$. Thus we need to have:

$$(*) \quad \text{Tr}_w(\gamma\delta[D,\delta]^4 D^{-4}) > 0.$$

But we already know how to find such a non-trivial pair $(\delta, D)$, since by the results of Section 4 we know that if $N$ is a compact spin 4-manifold and $\emptyset_N$ the Dirac operator on $N$, the value of $\psi_w$ on $C^\infty(N)$

is given by:

$$\psi_w(f^0,\ldots,f^4) = \frac{1}{8\pi^2}\int_N f^0\,df^1\wedge df^2\wedge df^3\wedge df^4 \quad \forall f^j \in C^\infty(N).$$

Thus to get $(*)$ one can take $D = \not{\partial}_N \otimes 1$, where 1 is the identity $q\times q$ matrix, so that $\mathfrak{H} = L^2(N,S) \otimes \mathbb{C}^q = L^2(N, S\otimes \mathbb{C}^q)$, and take for $\delta$ a smooth map $x \mapsto \delta(x)$ from $N$ to the Grassmannian $G_{q,r}$ of $r$-dimensional subspaces of $\mathbb{C}^q$ whose Chern classes $c_1$, $c_2$ satisfy $c_1^2 + c_2 > 0$. (It is a rather strange fact that in many examples where $(*)$ is computed, for a quadruple $(\mathfrak{H}, D, \gamma, \delta)$, its value turns out to be an integer.)

In our case we next need to extend the action of $\mathbb{C}\oplus\mathbb{C} \subset \mathcal{A}$ to $\mathcal{A}$ and thus the natural choice is to take $N = M$, so that the action of $\mathcal{A}$ is given by $a = (f, f') \mapsto f\delta + f'(1-\delta) \in \mathcal{L}(\mathfrak{H})$.

Thus let us fix $\delta \in C^\infty(M, G_{q,r})$ and consider the $4^+$ summable $K$-cycle on $\mathcal{A}$ given by:
1) $\mathfrak{H} = L^2(M, S\otimes \mathbb{C}^q)$, 2) $D = \not{\partial}_M \otimes 1$, 3) $\gamma = \gamma_M \otimes 1$,
4) $(f, f')\xi = (f\delta + f'(1-\delta))\xi$, $\forall \xi \in \mathfrak{H}$, $(f, f') \in \mathcal{A}$.

**Lemma 3** *The metric $d$ on $X = M \cup M$ associated to the $K$-cycle $(\mathfrak{H}, D, \gamma)$ coincides with the Riemannian metric $d_M$ on both copies of $M$ and for $x$, $y$ on different copies one has:*

$$d(x,y) = \inf_z(d_M(x,z) + d_M(z,y) + \ell(z)),$$

*where $\mu(z) = \ell(z)^{-1}$ is the norm of the differential $d\delta(z)$.*

If the function $\ell(z)$ varies slowly this metric is essentially the same as that of Model II, but for the two copies of $M$ in $X$ to be very close it is not at all necessary that $\mu(z)$ be large everywhere. In fact $\mu$ may very well vanish on a very large proportion of $M$; provided it takes large values $\mu(z_i)$ at some point in a neighbourhood of size $\epsilon$ of $x$ one will control $d(x, x')$ by $2\epsilon + \mu(z_i)^{-1}$.

The lemma follows from the equality:

$$i[D, a] = df\,\delta + (f - f')d\delta + df'(1-\delta), \quad \text{for } a = (f, f') \in \mathcal{A}.$$

We now proceed exactly as for Model II, and fixing the same finite projective module $\mathcal{E} = \begin{bmatrix} (1,1) & 0 \\ 0 & (0,1) \end{bmatrix} \mathcal{A}^2$ over $\mathcal{A}$ we compute the field content and the bosonic part of the action of Theorem 1 of Section 4. Since the conditions of Theorem 2 of Section 4 are fulfilled we know that the minimum of the action cannot be 0.

## 2. Essay on Physics and Non-commutative Geometry

As for Model II we let the connection $\nabla$ be of the form $\nabla \xi = e \, d\xi + \rho \xi$, where $\rho = -\rho^* \in M_2(\Omega^1(\mathcal{A}))$. We write $\rho = \begin{bmatrix} \rho_{11} & \rho_{12} \\ \rho_{21} & \rho_{22} \end{bmatrix}$ and $\rho_{kl} = \sum a_{klj} \, db_{klj}$, $a_{klj} = (f_{klj}, f'_{klj})$, $b_{klj} = (g_{klj}, g'_{klj})$.

Then the field content is as in Model II given by the fields:

$$
\begin{array}{ll}
A_1 = \sum f_{11j} \, dg_{11j}, & A'_1 = \sum f'_{11j} \, dg'_{11j}, \\
A_2 = \sum f'_{22j} \, dg'_{22j}, & W = \sum f'_{21j} \, dg'_{21j}, \\
\phi_1 = \sum f'_{11j}(g_{11j} - g'_{11j}), & \phi_2 = \sum f'_{21j}(g_{21j} - g'_{21j}).
\end{array}
$$

As in Model II there are additional fields such as the $\psi$'s and the two others given by:

$$ X = \sum f_{11j} \, dg'_{11j} \qquad Y = \sum f'_{21j} \, dg_{21j}. $$

However all these 6 fields will decouple and will be ignored by minimizing the action over them. If one takes them into account one gets the following formulae for the matrix elements of the curvature $\pi(\nabla^2) = \pi(e \, de \, de + e \, d\rho + \rho^2) = \theta = \begin{bmatrix} \theta_{11} & \theta_{12} \\ \theta_{21} & \theta_{22} \end{bmatrix}$:

$$
\begin{aligned}
\theta_{11} ={}& \delta(dA_1 + A_1^2 + \psi_1 - (\phi_1 + \bar\phi_1 + \phi_1\bar\phi_1 + \phi_2\bar\phi_2)(d\delta)^2) \\
& + [(-d\bar\phi_1 - (\bar\phi_1 + 1)A_1 + X)\delta \, d\delta \\
& + \delta \, d\delta(-(\bar\phi_1 + 1)A'_1 - \bar\phi_2 W + X)] + [\ \ ]^* \\
& + (1 - \delta)(dA'_1 + A'^2_1 + \psi'_1 - W^*W - (\phi_1 + \bar\phi_1 + \phi_1\bar\phi_1)(d\delta)^2), \\
\theta_{21} ={}& -\phi_2(\bar\phi_1 + 1)(1 - \delta)(d\delta)^2 + (1 - \delta)d\delta(\phi_2 A_1 - Y) \\
& + (d\phi_2 + W(1 + \phi_1) + A_2\phi_2 - Y)(1 - \delta)d\delta \\
& + (1 - \delta)(dW + WA'_1 + A_2 W + \psi), \\
\theta_{22} ={}& (1 - |\phi_2|^2)(1 - \delta)(d\delta)^2 + (dA_2 + A_2^2 - WW^* + \psi_2)(1 - \delta).
\end{aligned}
$$

One then checks that the component of degree 2 in the Clifford algebra of each $\theta_{ij}$ is independent of the six fields $\psi_1$, $\psi_2$, $\psi'_1$, $\psi$, $X$, $Y$ and that the freedom in the choice of these fields precisely cancels the component of degree 0 of $\theta_{ij}$. Thus for instance in the term:

$$ (-d\bar\phi_1 - (\bar\phi_1 + 1)A + X)\delta \, d\delta + \delta \, d\delta(-(\bar\phi_1 + 1)A'_1 - \bar\phi_2 W + X), $$

the 1-form $X$ disappears in the component of degree 2 and if one takes $-d\bar\phi_1 - (\bar\phi_1 + 1)A + X - (\bar\phi_1 + 1)A'_1 - \bar\phi_2 W + X = 0$ one cancels the component of degree 0.

Thus to compute our action we can just ignore the fields $\psi$, $X$, $Y$ and only consider the components of degree 2 of $\theta_{ij}$, in other words we are just working with matrices of ordinary differential forms. We then rewrite the relevant part of $\theta_{ij}$ as follows:

$$\begin{aligned}
\theta_{11} &= \delta(dA_1 - (|\phi_1 + 1|^2 + |\phi_2|^2 - 1)(d\delta)^2) \\
&\quad + (d\phi_1 + (\phi_1 + 1)(A_1' - A_1) - \phi_2 W^*)(1-\delta)d\delta + (\ )^* \\
&\quad + (1-\delta)(dA_1' - W^*W - (|\phi_1 + 1|^2 - 1)(d\delta)^2), \\
\theta_{21} &= -\phi_2(\bar\phi_1 + 1)(1-\delta)(d\delta)^2 \\
&\quad + (d\phi_2 + (\phi_1 + 1)W + \phi_2(A_2 - A_1))(1-\delta)d\delta \\
&\quad + (1-\delta)(dW + WA_1' + A_2 W) \\
\theta_{22} &= (1 - |\phi_2|^2)(1-\delta)(d\delta)^2 + (dA_2 - WW^*)(1-\delta).
\end{aligned}$$

We can, as for Model II, simplify these expressions using the connection $T = \begin{bmatrix} A_1' & -W^* \\ W & A_2 \end{bmatrix}$ and the covariant derivative $d_X\phi = d\phi + X\phi$, $X = T - A_1 1$, applied to the pair $\phi = (\phi_1 + 1, \phi_2)$. Thus our action is the sum $I(\rho) = |\theta_{11}|^2 + 2|\theta_{12}|^2 + |\theta_{22}|^2$ of the $L^2$ norms$^2$ of the $\theta_{ij}$. Each $\theta_{ij}$ is a $q \times q$ matrix of 2-forms and at each point $x \in M$ we are using the tensor product Hilbert space $\mathcal{H} = \Lambda^2 T_x^* \otimes M_q(\mathbb{C})$, where $M_q(\mathbb{C})$ is endowed with the Hilbert–Schmidt inner product. In such a tensor product the four subspaces $(1 \otimes \delta_x)\mathcal{H}(1 \otimes \delta_x)$, $(1 \otimes (1-\delta_x))\mathcal{H}(1 \otimes \delta_x)$, $(1 \otimes \delta_x)\mathcal{H}(1 \otimes (1-\delta_x))$ and $(1 \otimes (1-\delta_x))\mathcal{H}(1 \otimes (1-\delta_x))$ are pairwise orthogonal. This then allows us to express $I(\rho)$ using the curvature matrix $G = dT + T^2$ of the connection $T$, as follows:

$$\begin{aligned}
I(\rho) &= |\delta(dA_1 - (|\phi|^2 - 1)(d\delta)^2)|^2 \\
&\quad + |(1-\delta)(G_{11} - (|\phi^1|^2 - 1)(d\delta)^2|^2 \\
&\quad + 2|(1-\delta)(G_{21} - \phi^2 \bar\phi^1 (d\delta)^2)|^2 \\
&\quad + |(1-\delta)(G_{22} + (1 - |\phi^2|^2)(d\delta)^2|^2 \\
&\quad + 2|d_X\phi \wedge (1-\delta)d\delta|^2.
\end{aligned}$$

In order to have orthogonality of terms such as $(1-\delta)G_{21}$ and $\phi^2\bar\phi^1(d\delta)^2$ we just need to know that the two matrices of 2-forms:

$$K = \delta(d\delta)^2, \quad K' = (1-\delta)(d\delta)^2$$

are orthogonal to the subspace $\Lambda^2 T_x^* \otimes 1$ of $\mathcal{H}$ for every $x \in M$. This means that the 2-forms $\omega = \text{tr}(K)$, $\omega' = \text{tr}(K')$ should be identically 0, which we shall assume from now on.

We can now write down our action in simple terms:

$$I(\rho) = \int_M \mathcal{L},$$

where the Lagrangian $\mathcal{L}$ is:

$$\mathcal{L} = \dim(\delta)|dA_1|^2 + \dim(1-\delta)|G|^2 + 2|d_X\phi \wedge \Gamma|^2 \\ + (|\phi|^2 - 1)^2(|K|^2 + |K'|^2) + |K'|^2,$$

where $G$ is the field strength tensor of the connection $T$ and where $\Gamma = (1-\delta)d\delta$, $K = \delta(d\delta)^2$, $K' = (1-\delta)(d\delta)^2$. Note that $\Gamma^2 = 0$ and $|d\Gamma + \Gamma^2|^2 = |K|^2 + |K'|^2$.

We now have to face the question of the choice of $\delta$. This ought to be a very interesting problem in general relativity since it is this choice which specifies the geometric structure of our "double-sheeted" space-time $X = M \cup M$, besides the choice of the Riemannian metric on $M$. In fact, it is clear from the form of the Lagrangian $\mathcal{L}$ that the metric $g$ of $M$ only enters by its conformal class, the choice of volume element is thus dictated by the last term: $(|\phi|^2 - 1)^2(|K|^2 + |K'|^2)$, i.e. by the Yang–Mills norm of the field $\Gamma$. The only term which remains then is $|d_X\phi \wedge \Gamma|^2$ and to understand it we need to understand the Euclidean norm$^2$ on each cotangent space $T_x^*(M)$, given by:

$$(*) \quad \xi \mapsto |\xi \wedge \Gamma|^2.$$

There is no reason in general why this norm$^2$ should be related to the norm$^2$ $\xi \mapsto |\xi|^2$ since the choice of $\Gamma = (1-\delta)d\delta$ is *a priori* independent of the choice of the metric on $M$.

We shall just content ourselves with the following example where for symmetry reasons the two norms are proportional. Thus (cf. [1]) we let $M = S^4$ be the 4-sphere which we identify with $P_1(\mathsf{H})$ the quaternionic projective space, and we choose for $\delta$ the basic instanton, i.e. the map to the Grassmannian of quaternionic lines in $\mathsf{H}^2$, viewed as a subspace of the real Grassmannian $G_{4,8}$. It follows then from the natural $SO(5)$ equivalence of this situation that while the choice of this instanton uniquely specifies the volume form of $S^4$, i.e. specifies the metric in its conformal class, the norm$^2$ given above by $(*)$ is proportional to the metric. An extension of this to other manifolds ought to be linked with Penrose's twistor theory (cf. [1]).

Using the computations at the end of the discussion of Model II we thus see that provided $(*)$ is proportional to $|\xi|^2$ we do get exactly the Weinberg–Salam bosonic sector. There is however one number which in Model III plays the role of $\int_M \mu^4$ in Model II and hence ought to be quite large, it is: $\int_M |K|^2 + |K'|^2$, or better $\int_M (K')^2$, i.e. the instanton number, which is an integer. We shall leave the discussion of the relations between our Model III and general relativity, of its fermionic content, anomalies etc. to later investigations.

## References

[1] M.F. Atiyah, *Geometry of Yang-Mills Fields*, Academia dei Lincei, Pisa, 1979.

[2] H. Bacry, *The notion of space in quantum physics*, Lecture Notes in Physics **308**.

[3] J. Bellissard, Ordinary quantum Hall effect and non-commutative cohomology, in *Proc. of "Localization of disordered systems"*, Bad Schandau 1986, Teubner Publ., Leipzig, 1988.

[4] N.N. Bogoliubov and D.V. Shirkov, *Quantum Fields*, Benjamin, 1983.

[5] A. Connes and J. Cuntz, Quasi homomorphismes, cohomologie cycliques et positivité, *Commun. Math. Phys.* **114** (1988) 515-526.

[6] A. Connes, The action functional in non-commutative geometry, *Commun. Math. Phys.* **117** (1988) 673-683.

[7] A. Connes and M. Rieffel, Yang-Mills for non-commutative two tori, *Contemp. Math.* **62** (1987) 237-266.

[8] A. Connes, Non-commutative differential geometry, *Publ. Math. IHES* **62** (1985) 257-360.

[9] J. Dixmier, Existence de traces non normales, *C. R. Acad. Sci. Paris* **262** (1966) 1107-1108.

[10] M. Gromov, Groups of polynomial growth and expanding maps, *Publ. Math. IHES* **53** (1981) 53-78.

[11] M. Karoubi, Homologie cyclique et $K$ théorie, *Astérisque* **149** (1987).

[12] G. Mokobodski, Limites médiales, Exposé au Sem. Probabilités P.A. Mayer, Univ. de Strasbourg 1971-72, *Lect. Notes in Math.* **321** (1973) 198-204.

[13] M. Takesaki, *Theory of operator algebras*, Springer, New York-Heidelberg-Berlin.

[14] D. Voiculescu, Some results on norm ideal perturbations of Hilbert space operators, I. *J. Op. Theory* **1** (1979) 3-37; II. *ibid* **5** (1981) 77-100.

I.H.E.S., 35 ROUTE DE CHARTRES, 91440 BURES-SUR-YVETTE, FRANCE.

# 3

# Twistors, Particles, Strings and Links

*R. Penrose*

## 1. Basic twistor theory

Twistors provide a setting for physical theory that is different in several respects from the usual one. (See Huggett and Tod [7], Penrose and Rindler [14], [15], Ward and Wells [21] for further details.) In place of the conventional underlying geometry, namely a real four-dimensional space-time manifold, we have a complex manifold — whose complex dimensionality is either three (projective twistor space PT, with the structure of $\mathbb{CP}^3$) or four (non-projective twistor space T, with the vector space structure $\mathbb{C}^4$). The correspondence between space-time and twistor space is a non-local one, and questions of locality in space-time take on a new character when interpreted in terms of twistors. Instead of the normal description using space-time fields, which are essentially local objects (e.g. curvature, which can be of the Maxwell, Yang-Mills or Einstein variety), one has global information in twistor space: sheaf cohomology elements, in the case of the twistor description of linear fields, and global bundle structure or global manifold structure in the case of non-linear fields. A key ingredient of twistor theory is holomorphic (i.e. complex-analytic) structure. It is global holomorphic structure, in twistor theory, that replaces the local field equations of the normal space-time descriptions.

The most direct part of the twistor correspondence is the representation of entire light rays (null geodesics) in (conformally compactified) Minkowski 4-space M by single points in a five-real-dimensional submanifold PN of PT. The equation defining PN is

$$\Sigma(Z) := Z^\alpha \bar{Z}_\alpha = 0, \qquad (1)$$

where $\Sigma(Z)$ is a Hermitian-quadratic form (of signature $++--$) in terms of the twistor $Z \in \mathsf{T}$. (In the usual component description, $\bar{Z}_0, \bar{Z}_1, \bar{Z}_2, \bar{Z}_3$ are the complex conjugates of $Z^2, Z^3, Z^0, Z^1$ respectively.) The points of PT which do *not* lie on PN also have physical interpretations, namely as classical massless particles with varying energy and fixed (non-zero) spin, the *positive*-helicity particles being described by the points of $\mathsf{PT}^+$ (given by $\Sigma(Z) > 0$ and the *negative-helicity* ones, by the points of $\mathsf{PT}^-$ (given by $\Sigma(Z) < 0$). In fact, the points of $\mathsf{T}$ can themselves, up to a phase freedom

$$Z \mapsto e^{i\theta} Z, \tag{2}$$

be given a direct physical interpretation in terms of the momentum and angular momentum structure of classical massless particles (so the *modulus* of the scale factor for $Z$ is thereby interpreted). The 4-momentum and angular momentum have explicit expressions, bilinear in $Z^\alpha$ and $\bar{Z}_\alpha$. The helicity of the particle then becomes just the expression $\Sigma(Z)/2$. The phase of $Z$, also, can be interpreted (up to a sign) in terms of a polarization direction for the particle.

## 2. A generalization of conformal field theory

When we use the twistor coordinates $Z^\alpha$ and $\bar{Z}_\alpha$ in place of the usual space-time positions and momenta $x^a$, $p_a$ we find that space-time linear massless fields or, more appropriately, *massless single-particle wave functions*, have a concise description in terms of holomorphic sheaf cohomology of regions of $\mathsf{PT}$ (or of $\mathsf{T}$). A wave function for a massless particle of helicity $n/2$ is described in twistor terms by an element of the holomorphic sheaf cohomology group

$$H^1(\mathsf{PT}^+, \mathcal{O}(-n-2)). \tag{3}$$

Here $\mathcal{O}(k)$ denotes the sheaf of holomorphic functions "twisted by $k$" (i.e. described by holomorphic functions of $Z^\alpha$ homogeneous of degree $k$). The fact that the "domain" of the cohomology element is $\mathsf{PT}^+$ is what specifies the wave function as being of *positive frequency*. Elements of

$$H^1(\mathsf{PT}^-, \mathcal{O}(-n-2)). \tag{4}$$

define complex fields ("wave functions") of *negative frequency*. We can also define corresponding cohomology elements on the CR-manifold $\mathsf{PN}$, provided that we interpret this cohomology in the appropriate "CR-hyperfunctional" sense (cf. Andreotti and Hill [1], Penrose [11]), and we find, with this interpretation,

$$H^1(\mathsf{PN}, \mathcal{O}(-n-2)) = H^1(\mathsf{PT}^+, \mathcal{O}(-n-2)) \cup H^1(\mathsf{PT}^-, \mathcal{O}(-n-2)). \tag{5}$$

This is closely analogous to the splitting of complex-valued functions — or, more appropriately, "half-forms", given by $\mathcal{O}(-1)$ — into their positive and negative frequency parts according to whether they extend into the northern or southern hemisphere:

$$H^0(S^1, \mathcal{O}(-1)) = H^0(\mathbb{C}^+, \mathcal{O}(-1)) \cup H^0(\mathbb{C}^-, \mathcal{O}(-1)), \tag{6}$$

## 3. Twistors, Particles, Strings and Links

where $S^1$ denotes the equator ($= \mathbb{R} \cup \{\infty\}$) of the Riemann sphere, and $\mathbb{C}^+$ and $\mathbb{C}^-$ denote its (open) northern and southern hemispheres, respectively (and where for "equality" to hold in the above equation, rather than merely "inclusion", the first $H^0$ should be taken in the above hyperfunctional sense).

This analogy is suggestive of a way in which *conformal field theory* may be generalized to give a theory which could have direct relevance to interacting (massless) fields in four-dimensional space-time (Hodges, Penrose and Singer [6], Singer [19]). In place of the bounded Riemann surfaces of ordinary conformal field theory — i.e. generalizations of the region $\mathbb{C}^+ \cup S^1$ — we can consider three-complex-dimensional manifolds $X$ which generalize the region $\mathsf{PT}^+ \cup \mathsf{PN}$. Any such manifold $X$, which we call a *flat twistor space* (or, more colloquially, a "pretzel twistor space"), is a complex three-manifold which is compact with boundary, each component of the boundary being a copy of $\mathsf{PN}$ (i.e. with the same CR-structure as $\mathsf{PN}$). We may take $p$ of these $\mathsf{PN}$s to have positive orientation (representing in-states) and $q$ of them to have negative orientation (representing out-states). The manifold $X$ is to satisfy the requirement that each point in the interior of $X$ has a neighbourhood which is biholomorphic to a neighbourhood of a $\mathbb{CP}^1$ in $\mathbb{CP}^3$ (and, as a technical side condition, that the canonical bundle of $X$ admits a fourth root – so that the "non-projective" version of $X$ can be constructed).

The now-standard procedures of conformal field theory (see Segal [18] and Witten [22]) show how, given any Riemann surface, compact with a number of parametrized boundary $S^1$s some positively oriented and some negatively, one may construct an interacting quantum field theory, where each "in-state" is an element of a Fock space associated with a positively oriented $S^1$ and each "out-state", an element of a Fock space associated with a negatively oriented $S^1$. The twistor generalization carries this construction over to the three-complex-dimensional case virtually without change.

What we end up with, given a flat twistor space $X$, is (up to an undetermined phase factor) a multilinear map from a product of Fock spaces to $\mathbb{C}$:

$$\mathcal{H} \times \cdots \mathcal{H} \times \mathcal{H}^* \times \cdots \times \mathcal{H}^* \longrightarrow \mathbb{C}, \qquad (7)$$

where $p$ is the number of $\mathcal{H}$'s in the above, and $q$ is the number of $\mathcal{H}^*$'s. Here, $\mathcal{H}$ stands for the fermionic Fock space of free massless particles:

$$\mathcal{H} = \mathbb{C} \oplus H \oplus H \wedge H \oplus H \wedge H \wedge H \oplus H \wedge H \wedge H \wedge H \oplus \cdots, \qquad (8)$$

where each element of $H$ describes a single-particle state for a massless particle (or anti-particle). Here the helicity of the massless particle need not be specified, so all helicities are considered together. (On the other hand, we could choose to fix the spin to be $n/2$, but then we would have to replace $\mathcal{O}$ by $\mathcal{O}(-n-2) \oplus \mathcal{O}(n-2)$ in what follows.) Note that, as things stand, the Fock space is "fermionic" irrespective of whether the helicity is integral or half-odd integral. I shall have a comment to make about this in a moment.

The twistor description of a single-particle massless state is an element of $H^1(\mathbb{N}, \mathcal{O})$ of positive frequency, i.e. an element of $H^1(\mathsf{T}^+, \mathcal{O})$. Thus we can take
$$H = H^1(\mathsf{T}^+, \mathcal{O}), \tag{9}$$
but this ignores the usual quantum-mechanical requirement that states be normalizable. In twistor terms, this normalizability condition is a somewhat awkward "$L^2$" condition on the cohomology elements. A stronger condition ensuring normalizability (but not completeness of the space of states), and which is sufficient for many purposes, is that the "domain" of the cohomology element extend to $\mathsf{T}^+ - \{0\}$, i.e. to that part of $\mathsf{T}$ which lies above the closure $\overline{\mathsf{PT}^+}$, of $\mathsf{PT}^+$. This is equivalent to requiring that the massless fields, in their space-time description, be real-analytic (including at infinity — so that, for example, momentum states are excluded). This would be the analogue, in ordinary conformal field theory, of a requirement that the functions on $S^1$ be taken to be real-analytic.

An essential feature of our generalization of conformal field theory to complex manifolds of higher dimension is that quanta are represented by $H^1$ cohomology elements, rather than just by ordinary functions (i.e. $H^0$ cohomology elements). In standard twistor theory, the fact that field quantities are represented by $H^1$s rather than by $H^0$s has important implications since $H^1$ elements can often represent infinitesimal deformations of complex structure. When such infinitesimal deformations are "exponentiated" to become finite deformations, it is sometimes possible to interpret the resulting deformed twistor spaces at providing the solutions of *non*-linear fields in space-time. The (anti-)self-dual Einstein and Yang-Mills fields have been successfully described in this way a number of years ago (Penrose [10], Ward [20]). It is to be expected that this feature will have an important role to play in the present theory also. (Some comments in relation to general relativity will be made shortly.)

## 3. Interpretation of Fock space elements

As with ordinary conformal field theory, there is a problem of how one is to interpret the various "many-particle" elements of the Fock space. What we are presented with is an amplitude (up to an overall phase factor) for each specification of $p$ in-states and $q$ out-states. We would like to think of the in-states as representing incoming particles and the out-states as outgoing particles, but since the in-states represent elements of an entire Fock space, and so do the out-states, we appear to have the possibility of a several-particle state for *each* individual in- or out-state, rather than just a one-particle state for each. In string theory, one takes the "one-particle" elements of the Fock space to represent a lowest mode of excitation for a single particle, and the "many-particle" elements to represent its higher modes of excitation, rather than representing several actual particles. It is proposed that we do something similar here, but now we are able to interpret these "higher modes" by taking advantage of the ideas of twistor particle theory (see Penrose [9], Perjés [16], [17], Hughston [8]).

In twistor particle theory, massless particles (of the normal type, where the Pauli-Lubański spin 4-vector is taken to be proportional to the momentum 4-vector) are described by holomorphic functions of a single twistor variable $Z$. Actually any such function — referred to as a "twistor function" $f(Z)$ — is a really just a representative function for a sheaf cohomology element, namely the element of $H = H^1(\mathsf{T}^+, \mathcal{O})$ that represents the particle's wave function, as referred to above. But loosely we can refer to $f(Z)$ as the twistor description of the massless particle's wave function. To get from the twistor function to the particle's space-time wave function $\phi$ we apply a contour integration, which we can write symbolically as

$$\phi = \oint f \qquad (10)$$

(see Penrose and Rindler [15], Huggett and Tod [7] for details). How do we describe *massive* particles in twistor theory? Again we can use a twistor function, but now it must be a function of *several* twistor variables, e.g.

$$F(Y, Z) \quad \text{or} \quad G(X, Y, Z). \qquad (11)$$

There are contour integral expressions for the space-time "field" obtained from the twistor function just as before, but (as of now) the correct cohomological interpretation of $F$, $G$ etc. remains elusive. (The likelihood seems to be that $F$ and $G$ describe elements of *relative* sheaf cohomology, rather than of just ordinary sheaf cohomology.) For twistor functions of a single variable there is just one (scalar) quantum number, namely the *helicity*, and this is described twistorially in terms of the *homogeneity degree* of the twistor function (i.e. the helicity value $n/2$ arising from $\mathcal{O}(-n-2)$). For twistor functions of two variables, there are four scalar quantum numbers, two of which are the spin and mass. The remaining two can be given tentative interpretations in terms of electric charge and lepton number. For twistor functions of three variables, there are eight scalar quantum numbers, two of which are again spin and mass, and the remaining six of which can be tentatively interpreted in terms of hadronic $SU(3)$ quantum numbers. For more twistor variables we get still further quantum numbers.

The idea, then, is to interpret the "many-particle" elements of the Fock space as describing higher order "excitations of twistor space" which would correspond to single *massive* particles: leptons, $SU(3)$ hadrons, "exotic" hadrons, etc. Many puzzles remain to be sorted out concerning this. Not least among these is the fact that conformal invariance (for the space-time fields) is lost once we pass to the descriptions of massive particles. In twistor terms, this means that, for these higher excitations, the full symmetry of the space PN is lost, and a particular line I in PN representing infinity in Minkowski space M, must be singled out.

## 4. Gravitational interactions

The question of how gravity and space-time curvature is to be represented in such a twistorial scheme also presents difficult problems. It is to be expected that the "flat" twistor spaces defined above will have to be generalized to become "curved" in some appropriate sense. The asymptotic twistor spaces of asymptotically flat space times (cf. Penrose and MacCallum [13]) are relevant models; but so far these provide appropriate descriptions of gravitons of one helicity only (the left-handed helicity, according to the usual twistorial conventions). There are also ways of incorporating the information of the opposite helicity into the structure of such a twistor space, but I do not think that these are yet fully satisfactory. (A fully appropriate description would require an adequate solution of what has become known as the "googly problem": the problem of incorporating the information of right-spinning gravitons in the structure of a twistor space, whose ordinary deformations encode only left-spinning gravitons.) Again, there must be a violation of conformal invariance in the deformations that are allowed.

It should be remarked that despite many grandiose claims that the "conventional" (super)string theories can provide us with what is referred to as a "consistent quantum gravity theory", such schemes do not at all come to terms with most of the profound issues that are involved in bringing together the principles of quantum theory and those of curved space-time geometry. Curved space-time is not really an integrated feature of the (super)string models that have been seriously proposed so far. All the detailed descriptions refer only to perturbation expansions about a flat space background, where many of the essential issues of principle are not even addressed. Moreover, the terms of the perturbation expansion involve only "linear gravitons" (spin-two massless quanta in flat space-time M). There are good reasons for believing that such quanta do not give the correct descriptions for what would be the appropriate quantization of Einstein's general relativity. In Einstein's theory, gravitational radiation modes do not really decouple at infinity. Any burst of outgoing radiation will affect the very geometry of future null infinity, and so will influence the description of any burst of radiation that may be emitted later. This has serious implications for any theory of quantum gravity that is merely Poincaré invariant, and does not take into account behaviour under the BMS (Bondi-Metzner-Sachs) group (cf. Penrose and Rindler [15], Section 9.8). Related to this are certain infra-red problems, which are also intimately related to the BMS group (Ashtekar [2]; cf. also Penrose [10] for some different objections to the use of linear gravitons for a description of quantum gravity). The present proposal for a twistorial generalization of conformal field theory in principle allows genuine space-time curvature to be incorporated.

## 5. Twistor diagrams

Another important motivation for introducing this generalization is that it offers some hope for a systematic development of *twistor diagram theory* . This would be analogous to the way that string theories and conformal field theories provide a systematic and consistent way of describing scattering amplitudes where many Feynman diagrams would otherwise have to be summed up with ambiguities arising from "duality" equivalences. Twistor diagrams are the analogues, in twistor theory, of the Feynman diagrams of conventional theory. They have been systematically developed by Hodges ([3],[4],[5]), and can (at least in principle) provide an alternative to Feynman diagrams for computing amplitudes for scattering processes in quantum electrodynamics or standard electroweak theory. However, the procedures so far are rather more "ad hoc" than one would wish, and one has no entirely "twistorial" rule, as yet, for determining which twistor diagrams must be summed in a given scattering process. The hope is that our twistorial generalization of conformal field theory will eventually lead to such a rule, rather in the way that "duality diagrams" and the Riemann surfaces of string theory provide a consistent and systematic procedure for computing amplitudes.

## 6. Holomorphic linking

So far, not a great deal of progress has been made in finding a direct relationship between twistor diagrams and the building up of different kinds of pretzel twistor space (though it is possible to see certain basic connections). On the other hand, there is another set of ideas that have become popular amongst mathematical physicists, and which appear to have some significant relationship with twistor diagram theory. This is the theory of braids, knots and links, and there is some reason to expect that these ideas may ultimately tie in with the procedures for building up the different kinds of pretzel twistor spaces.

To understand why twistor theory might have relevance to knot theory, etc., consider the concept of a topological quantum field theory which, as Witten has shown, can provide fresh insights into the new knot and link invariants of Jones and others. The central feature of such field theories is the fact that there is *no local field information* (e.g. with the the case of "2+1-dimensional general relativity", as studied by Witten [22]), and everything of interest lies in the global structure. This is, of course, just what is wanted from the point of view of mathematical applications, these being of an essentially topological character. However, if we anticipate topological field theories having a direct relevance to *physics*, we are presented with a puzzle, since much physical information is normally described in terms of *local* field quantities, such as Einstein, Maxwell, or Yang-Mills curvature. This puzzle finds a natural resolution if we take the background space of the topological quantum field theory to be *twistor space* rather than space-time. As we have seen above, physical field information is coded *non-locally* in twistor space, and all local information actually disappears, as would be required in a topological theory. Accordingly, twistor space might indeed be more physically natural than space-time, as the background for a topological quantum field theory.

There is an immediate difficulty, however, in that twistor space is a *complex* space. If we try to think of what should be the complex analogue of a knot or a link in ordinary three-dimensional space, we are presented with a problem. A complex curve is *two*-dimensional, as a real manifold, and the ambient complex three-dimensional space would be *six*-dimensional, as a real manifold. Two-dimensional manifolds do not knot or link in six-space. However, with regard to linking, at least, there is the remarkable expression due to Gauss:

$$L(\mathcal{X}, \mathcal{Y}) = (4\pi)^{-1} \oint_\Gamma \{((\mathbf{x} - \mathbf{y}) \cdot (\mathbf{x} - \mathbf{y})\}^{-3/2}(\mathbf{x} - \mathbf{y}) \cdot d\mathbf{x} \wedge d\mathbf{y},$$

where $\mathbf{x}$ and $\mathbf{y}$ are the position vectors (in Euclidean space $\mathbb{E}^3$) of independent variable points on two closed curves $\mathcal{X}$ and $\mathcal{Y}$, respectively, and $L(\mathcal{X}, \mathcal{Y})$ provides the linking number of the two curves. We can interpret this integral expression perfectly well when the curves $\mathcal{X}$ and

$\mathcal{Y}$ become complex and lie in an ambient complex Euclidean space $CE^3$ and where the integration is over some two-dimensional (real) contour $\Gamma$ in the product space of the complex curves $\mathcal{X}$ and $\mathcal{Y}$. Indeed, since we are concerned specifically with twistor space here, we shall regard $CE^3$ to be embedded in a $CP^3$ and we identify this $CP^3$ with PT.

The complex curves themselves need not be compact, but the real contour $\Gamma$ is taken to be compact. Normally, $\Gamma$ would be assumed to have topology $S^1 \times S^1$ with one $S^1$ on $\mathcal{X}$ and one $S^1$ on $\mathcal{Y}$, but the expression would also make sense with more general choices for $\Gamma$. The integral is well defined provided that $\Gamma$ avoids the "zero-distance locus" $\mathcal{Z}$, given by the vanishing of the expression $(\mathbf{x} - \mathbf{y}) \cdot (\mathbf{x} - \mathbf{y})$ (although there is also a possible difficulty in making sure that the square root of this expression can be taken consistently over the whole of $\Gamma$). The value of $L(\mathcal{X}, \mathcal{Y})$ depends upon the choice of an element of second homology, namely that represented by $\Gamma$, in the space

$$\mathcal{X} \times \mathcal{Y} - \mathcal{Z}.$$

However, the actual location of $\mathcal{Z}$ within $\mathcal{X} \times \mathcal{Y}$ is dependent upon a specific selection of (complexified) Euclidean metric on $CE^3$. This has a certain arbitrariness from the point of view of $CP^3$, involving selecting a particular "absolute conic" $\Omega$ lying in $CP^3 - CE^3$. In fact the value of $L(\mathcal{X}, \mathcal{Y})$ is not very sensitive to this arbitrariness, and any continuous variation of $\Omega$ which does not result in $\mathcal{Z}$ meeting $\Gamma$ will leave it unchanged. The exact kind of "object" that is referred to by $L(\mathcal{X}, \mathcal{Y})$ still remains somewhat mysterious, and more subtle issues are involved than just the "linking" that occurs in the real case.

These matters are currently being investigated (cf. Penrose [12]). Certain significant interrelationships with twistor diagram theory have emerged, but this is not the place to elaborate upon these developments. Various different-looking contour integral formulae arise, and it is to be hoped that (twistor-motivated) expressions of this kind may lead to some new insights concerning links and knots, quantum field theory, and the relationship between such matters and twistor theory.

## References

[1] A. Andreotti and C.D. Hill, E.E Levi convexity and the Hans Lewy problem. Part I: Reduction to vanishing theorems; Part II Vanishing theorems, *Ann. Scuola Norm. Sup. Pisa Cl. Sci.* (**3**)**26** (1972) 747–806.

[2] A. Ashtekar, Quantization of the Radiative Modes of the Gravitational Field, in *Quantum Gravity 2*, ed. C.J. Isham, R. Penrose and D.W. Sciama, Clarendon Press, Oxford, 1981, p. 416–438.

[3] A.P. Hodges, Twistor diagrams, *Physica* **114A** (1982) 157–75.

[4] A.P. Hodges, A twistor approach to the regularization of divergences, *Proc. Roy. Soc. London* **A397** (1985) 341-74; Mass eigenstates in twistor theory, *ibid*, 375-96.

[5] A.P. Hodges, 1989, in *this volume*.

[6] A.P. Hodges, R. Penrose and M.A. Singer, A twistor conformal field theory for four space-time dimensions, *Phys. Lett.* **B216** (1989) 48-52.

[7] S.A. Huggett, and K.P. Tod, *An Introduction to Twistor Theory*, London Math. Soc. Student Texts, L.M.S. Publ.,1985.

[8] L.P. Hughston, *Twistors and Particles*, Lecture Notes in Physics No. 97, Springer-Verlag, Berlin, 1979.

[9] R. Penrose, Twistors and particles: an outline, in *Quantum Theory and the Structures of Time and Space*, eds. L. Castell, M. Drieschner and C.F. von Weizsäcker, Carl Hanser Verlag, Munich, 1975.

[10] R. Penrose, Non-Linear gravitons and curved twistor theory, *Gen. Rel. Grav.* **7** (1976) 31-52.

[11] R. Penrose, Physical space-time and non-realizable CR-structures, *Proc. Symp. Pure Math*, **39** (1983) 401-22.

[12] R. Penrose, Topological QFT and Twistors: Holomorphic Linking, *Twistor Newsletter* **27**, 1988; Holomorphic Linking: Postcript, *ibid*.

[13] R. Penrose and M.A.H. MacCallum, Twistor theory: an approach to the quantization of fields and space-time, *Phys. Rept.* **6C** (1972) 241-315.

[14] R. Penrose and W. Rindler, *Spinors and Space-Time, Vol. 1: Two-Spinor Calculus and Relativistic Fields*, Cambridge University Press, Cambridge, 1984.

[15] R. Penrose and W. Rindler, *Spinors and Space-Time, Vol. 2: Spinor and Twistor Methods in Space-Time Geometry*, Cambridge University Press, Cambridge, 1986.

[16] Z. Perjés, Perspectives of Penrose theory in particle physics *Rept. Math Phys.* **12** (1977) 193-211.

[17] Z. Perjés, Introduction to twistor particle theory, in *Twistor Geometry and Non-Linear Systems*, ed. H.D. Doebner and T.D. Palev, Springer-Verlag, Berlin, 1982, pp. 53-72.

[18] G.B. Segal, The definition of conformal field theory, 1989, *to appear*.

[19] M.A. Singer, in *this volume*.

[20] R.S. Ward, On self-dual gauge fields, *Phys. Lett.* **61A** (1977) 81-2.

[21] R.S. Ward and R.O. Wells, Jr., *Twistor Geometry*, Cambridge University Press, Cambridge, 1989, *to appear*.

[22] E. Witten, Quantum field theory, grassmannians and algebraic curves, *Commun. Math. Phys.* **113** (1988) 529-600; 2+1 dimensional gravity as an exactly solvable system, *Nucl. Phys.* **B311** (1988/89) 46-78.

MATHEMATICAL INSTITUTE, 24-29 ST. GILES', OXFORD OX1 3LB, ENGLAND.

# 4

# Instantons in Yang–Mills theory

S.K. Donaldson

In this note we consider some of the uses of the "instanton" solutions of the Yang-Mills equations. The discussion will touch briefly on ideas from a number of different areas, ranging from quantum field theory to enumerative algebraic geometry. Our account will be very sketchy (and even inaccurate on some technicalities) at a number of points; our aim is more to raise some questions which may be useful goals for research, rather than to present detailed or rigorous results.

## 1. Yang–Mills instantons

We recall briefly the main features of Yang-Mills theory over 4-dimensional manifolds. Let $P$ be a bundle over a compact oriented Riemannian 4-manifold $X$, with structure group $SU(2)$ (the last assumption being made primarily for simplicity). We suppose the second Chern class of $P$ is a positive integer $k$. For a connection $A$ on $P$ the Yang-Mills action $S(A)$ is

$$S(A) = \int_X |F(A)|^2 d\mu. \tag{1}$$

The instantons are, by definition, the connections satisfying the anti-self-dual equation, that the curvature be a purely anti-self-dual form (with respect to the Hodge $*$ operator defined by the metric on $X$). These are absolute minima of the action: $S(A) \geq 8\pi^2 k$ with equality if and only if $A$ is anti-self-dual. The instanton equation is a non-linear first order partial differential equation for the connection and, when allowance is made for the action of the gauge symmetry group $\mathcal{G}$, it is an elliptic equation. The space of solutions, modulo gauge equivalence, is a finite dimensional space; the moduli space $M$ (or $M_k$).

These instantons were first noted in the context of quantum field theory. The partition function $Z(t)$ of the Yang-Mills system is given, at least formally, by the integral over the space $\mathcal{A}$ of all connections:

$$Z(t) = \int e^{-tS(A)} \mathcal{D}\mathcal{A}. \tag{2}$$

When the parameter $t$ is small one expects the dominant contribution to the integral to come from the connections $A$ with $S(A)$ close to its minimal value; that is, from connections close to the instantons. More precisely the asymptotic behaviour of $Z(t)$ should, according to the precepts of quantum field theory, be modelled on the integral over the finite dimensional moduli space :

$$\int_M \sqrt{\frac{\det \Delta_0}{\det \Delta_+}} d\nu. \tag{3}$$

Here $\Delta_0, \Delta_+$ are respectively the Laplace operators on 0-forms and self dual 2-forms coupled to the connection, and the determinants are renormalised by zeta function methods.

In principle a good understanding of the instanton solutions would enable one to evaluate the integrals (3) over the moduli spaces, or at least to get qualitative information about them. The renormalised determinants appearing in the integrand are still rather exotic for mathematicians, but the other ingredient in the formula– the measure $d\nu$ induced by the natural Riemannian structure on the moduli space – is more accessible. At least as a beginning then it is interesting to understand the qualitative properties of this Riemannian metric.

**Question 1** *Do the Yang-Mills moduli spaces have (a) finite diameter, (b) finite volume?*

Substantial progress in understanding these metrics has been made recently by Groisser and Parker [9], see also [6]. It seems very likely that the answer to both parts of Question 1 is affirmative, at least for all practical purposes. The moduli spaces are typically incomplete but their metric completions coincide with the natural compactifications introduced in [3]. The key to the proofs is the convergence theorem of Uhlenbeck for instantons. Refining Question 1 one could ask for concrete bounds on the diameter and volume of the moduli spaces; how, for example, do they grow with the Chern class $k$? On rather flimsy evidence we might propose:

**Conjecture 2** *For a fixed metric on the base manifold $X$ the diameter of the moduli space $M_k$ is $O(k^{1/2})$ and the volume is $O(k^{\dim M_k})$.*

Any bound would be interesting, the search for volume and diameter bounds can be regarded as a search for effective bounds in the analytical arguments used in Uhlenbeck's work.

## 2. Differential topological invariants

More recently additional motivation for the study of instanton moduli spaces has come from the field of differential topology. The moduli spaces can be used to define invariants of the underlying smooth 4-manifold $X$. In outline these are defined by pairing the fundamental class $[M]$ of the moduli space with cohomology classes over the space of all gauge equivalence classes of (irreducible) connections $\mathcal{A}^*/\mathcal{G}$. The key point is that if we consider two different metrics $g_0$, $g_1$ on the base manifold $X$ the moduli spaces $M(g_0)$, $M(g_1)$ differ by a boundary in $\mathcal{A}^*/\mathcal{G}$, so the homological pairing should be independent of the choice of metric, and hence yields the desired invariant of $X$. The cohomology of $\mathcal{A}^*/\mathcal{G}$ (at least rationally) is well understood. One obtains cohomology classes from homology classes on $X$ by the following procedure. There is a universal bundle $\mathbf{E}$ over the product $X \times \mathcal{A}^*/\mathcal{G}$, and this has a Chern class $c = c_2(\mathbf{E})$ in $H^4(X \times \mathcal{A}^*/\mathcal{G})$. Now for any class $\alpha$ in $H_*(X)$ we can take the slant product $\alpha \backslash c$ to get a cohomology class $\mu(\alpha)$ over $\mathcal{A}^*/\mathcal{G}$. (The slant product is effectively the Kunneth decomposition of $H^*(X \times \mathcal{A}^*/\mathcal{G})$). We can then take cup products in the cohomology ring to generate all the rational cohomology of $\mathcal{A}^*/\mathcal{G}$. Thus our invariants can be viewed as multilinear functions on the homology of the underlying space $X$:

$$q(\alpha_1, \ldots, \alpha_r) = \langle \mu(\alpha_1) \ldots \mu(\alpha_r), [M] \rangle. \tag{4}$$

Here $\alpha_i$ is a homology class of degree $e_i$, say, and $\sum_1^r (4 - e_i) = \dim M_k$. The dimension of $M_k$ is given by the index formula $\dim M_k = 8k - 3(1 - b_1(X) + b_2^+(X))$, where $b_1$ is the first Betti number and $b_2^+$ is the "positive part" of the second Betti number, defined by the intersection form of $X$. If the moduli space has even dimension $2d$ and $\alpha$ is a two-dimensional homology class, we write

$$\gamma(\alpha) = q(\underbrace{\alpha, \ldots, \alpha}_{d \text{ times}}).$$

To make a precise and rigorous definition of these invariants one needs to get around the non-compactness of the moduli spaces. This is done by choosing particular representatives of the cohomology classes over $\mathcal{A}^*/\mathcal{G}$ so that their support meets the moduli spaces in a compact set [5]. However there are other natural representatives which should be able to serve the same purpose. Suppose in particular that $\alpha$ is a 2-dimensional homology class on $X$ whose Poincaré dual can be represented in de Rham cohomology by a self-dual 2-form $\theta$ on $X$. Then the restriction of $\mu(\alpha)$ to the moduli space is represented by the 2-form $\Theta$ :

$$\Theta(a,b) = \int_X Tr(a \wedge b) \wedge \theta. \tag{5}$$

Here $a$ and $b$ are tangent vectors in the moduli space, i.e. bundle valued 1-forms over $X$. If the moduli space has even dimension $2d$ the invariant $\gamma(\alpha)$ is given formally by the integral :

$$\gamma(\alpha) = \int_M \Theta^d. \tag{6}$$

**Problem 3** *Prove that the integral (6) converges and its value coincides with the invariant defined in [5].*

Modulo the solution of Problem 3 we can say that quantum field theory and differential topology lead to broadly similar questions, in that both seek to evaluate integrals over the moduli spaces of instantons, in the one case of functions, formed from renormalised determinants, with respect to the Riemannian measure and in the other case of natural differential forms representing de Rham cohomology classes.

## 3. Kahler manifolds

In the discussion above the 2-form $\Theta$ on the moduli space is uniformly bounded with respect to the Riemannian metric, so $\Theta^d$ is bounded with respect to the volume element and, modulo the solution to Problem 3, we get an inequality:

$$\gamma(\alpha) \leq d! \|\frac{1}{\sqrt{2}}\theta\|^d \text{volume}(M), \tag{7}$$

where $\alpha$ is the Poincaré dual of the de Rham class defined by a self-dual 2-form $\theta$, and the norm is the supremum norm over $X$. This gives additional motivation for Conjecture 2, since volume bounds on the moduli spaces would lead to information about the topological invariants. In the case when $X$ is a complex Kahler surface we get a sharper relation. We can then take $\theta$ to be the Kahler form $\omega$ on $X$. It turns out that the moduli space $M$ is then naturally a complex manifold, its Riemannian metric is again Kahler and the Kahler form $\Omega$ on $M$ is precisely the one associated to $\omega$ by the formula (5). So (again modulo Problem 3) we have the equality :

$$\gamma(\alpha) = d! \text{ volume}(M), \tag{8}$$

for the class $\alpha$ dual to $\omega$.

The complex structure on the moduli space in this situation is a reflection of a fundamental link between the instanton equations and holomorphic geometry. For an instanton $A$ we can form the coupled Cauchy-Riemann operator on the associated complex vector bundle $E$ over $X$. The instanton equations imply that the coupled Cauchy-Riemann equations are integrable and the solutions endow $E$ with the structure of a holomorphic bundle. Conversely for any holomorphic bundle which satisfies a certain non-degeneracy condition known to algebraic geometers as "stability" one can recover an instanton. So the moduli spaces of instantons can be identified with moduli spaces of stable holomorphic bundles. The theory involved in this relation is quite rich. In particular renormalised determinants enter the picture in a natural way. The instanton corresponding to a holomorphic bundle can be obtained by minimising over all compatible connections on the bundle an expression formed from a combination of determinants. In turn the minimal value gives, roughly speaking, a function on the moduli space which is a Kahler potential for the metric form $\Omega$. More precisely these determinants define a connection on a determinant line bundle $\mathcal{L}$ over the moduli space which has curvature proportional to $\Omega$. So again one sees a close affinity between the ideas coming from quantum field theory and those from geometry and topology.

## 4. Witten's topological quantum field theory

Another link between the quantum field theorists' and differential topologists' study of instantons has been forged in the last year by Witten [11]. He shows that, at least formally, one can define a quantum field theory whose partition function, of the form :

$$W(t) = \int\int e^{-tS(A,\phi)} \mathcal{D}A\mathcal{D}\phi, \qquad (9)$$

is independent of $t$ and yields a (differential) topological invariant of the underlying 4-manifold $X$. Here the integral runs over a space of pairs $(A,\phi)$ where $A$ is a connection and $\phi$ is an auxiliary field. Carrying out the $\phi$ integration first we can write $W(t)$ as an integral over $\mathcal{A}$. As the parameter $t$ tends to 0 the integrand localises around the instanton moduli space and Witten shows that we recover the invariants described in Section 2 above. Atiyah has shown how one can interpret Witten's construction as "de Rham cohomology in the infinite dimensional space $\mathcal{A}/\mathcal{G}$", or better, equivariant de Rham cohomology on $\mathcal{A}$ [1]. This formulation is very natural and generalises familiar constructions in finite dimensional topology. In the simplest example, consider a proper map $f$ from $\mathbf{R}^n$ to $\mathbf{R}^n$. Such a map has an integer invariant, the degree, which can be defined either by:

(a) Counting, with signs, the points in the preimage $f^{-1}(y)$ of a regular value $y$ of $f$. (The counting should be regarded as evaluating the canonical 0-dimensional cohomology class on the 0-cycle carried by $f^{-1}(y)$).

(b) Choose an $n$-form $\rho$ on $\mathbf{R}^n$, of integral 1, and evaluate the integral of the pull-back $f^*(\rho)$ over $\mathbf{R}^n$.

To reconcile the two definitions one can take a family of forms $\rho_t$, with support localising at $y$ as $t$ tends to 0. This familiar correspondence is plainly similar in style to that in Witten's work. A more precise analogue is to consider a vector bundle $V$ of rank $n$ over an $n+r$ dimensional manifold $N$. The appropriate topological invariant is then the Euler class in $H^n(N)$. If $s$ is a generic section of $V$ the zero set $Z$ of $s$ is an $r$-dimensional submanifold of $N$, Poincaré dual to the Euler class. On the other hand, in the presence of suitable differential geometric structures on $V$ and $N$, we can represent the Euler class by explicit differential forms defined by the local geometry. Atiyah shows that Witten's construction gives, in this general setting, new representatives for the Euler class, closely related to the work of Mathai and Quillen [10]. On the other hand the more elementary definition of the invariants uses the moduli space, analogous to the zero set $Z$ above. In the Yang-Mills picture $N$ becomes the space $\mathcal{A}^*/\mathcal{G}$, $V$ becomes the vector bundle $\mathcal{V}$ over $\mathcal{A}^*/\mathcal{G}$ with fibres isomorphic to $E = \Omega^2_+(\mathrm{ad}P)$, associated to the principal $\mathcal{G}$ bundle $\mathcal{A}^* \to \mathcal{A}^*/\mathcal{G}$ by the natural action of $\mathcal{G}$ on $E$. The section $s$ of $\mathcal{V}$, cutting out $M$ as its zero set, is defined by the $\mathcal{G}$-equivariant map $A \mapsto F(A) + *F(A)$. Formally then our invariants are defined by the Euler class of the infinite dimensional bundle $\mathcal{V}$ over $\mathcal{A}^*/\mathcal{G}$.

## 5. Calculation of invariants, comparison with two dimensional case

For the remainder of this paper we will address the problem of calculating the invariants defined by instanton moduli spaces, as outlined in Section 2. We concentrate on the case when the base space $X$ is a complex Kahler manifold, in fact a complex algebraic surface (i.e. 2 complex dimensions), since we can then appeal to the correspondence described in Section 3 and describe the moduli spaces in terms of holomorphic bundles. The problem we have then is to calculate the pairings (4) between the moduli spaces of stable bundles over a complex algebraic surface and products of the cohomology classes constructed from the universal bundle. This problem has an analogue in two (real) dimensions which, while it does not have the same significance for the definition of new invariants, appears to go beyond our present knowledge and is related to various other developments, notably in 2-dimensional conformal field theory. It seems likely that it is important to understand this 2-dimensional case, since it is in a

## 4. Instantons in Yang–Mills theory

sense embedded in the higher dimensional problem (for example we can consider bundles over a product of Riemann surfaces). For this analogous problem we consider Yang-Mills theory over a compact Riemann surface $\Sigma$ of genus $g$. The relation with holomorphic bundles is now simpler— any connection defines a holomorphic structure. We again have a finite dimensional moduli space $W = W(\Sigma)$ which has two descriptions, either as moduli spaces of stable holomorphic bundles or as moduli spaces of flat unitary connections. The key theorem here, the prototype of that described in Section 3, is the result of Narasimhan and Seshadri asserting that these descriptions are indeed equivalent. The formulation in terms of Yang-Mills theory, together with many applications, was developed by Atiyah and Bott in [2]. In this 2-dimensional case we again look at the space of connections $\mathcal{A}$ over $\Sigma$, acted on by the gauge group $\mathcal{G}$. We have a universal bundle over $\mathcal{A}^*/\mathcal{G}$, and hence a slant product map :

$$\mu : H_*(\Sigma) \longrightarrow H^*(\mathcal{A}^*/\mathcal{G}). \qquad (10)$$

We define a ring $R$ to be the free graded-commutative ring generated by the homology of $\Sigma$, the tensor product of an exterior algebra on generators $\phi_\lambda$ in dimension 3 corresponding to a basis of $H_1(\Sigma)$ and a polynomial algebra on generators $\Omega, u$ in dimensions 2 and 4 respectively, associated to canonical generators of $H_2(\Sigma)$, $H_0(\Sigma)$. Taking cup products we get a map $M$ from the ring $R$ to the cohomology ring of $W$. The analogous problem is to identify the evaluation map $e : R_{6g-6} \longrightarrow \mathbf{Z}$ defined by :

$$e(P(\phi_\lambda, \Omega, u)) = \langle M(P(\phi_\lambda, \Omega, u)), [W] \rangle. \qquad (11)$$

(We note that $W$ has dimension $6g-6$. Actually there are complications if we consider $SU(2)$ connections here since the moduli space will not then be compact, due to reducible connections, a facet we have ignored hitherto in this paper. However the theory goes through smoothly for the analogous moduli spaces of $SO(3)$ connections, with non-zero Stiefel Whitney class $w_2$, which is compact.)

**Problem 4** *Identify the evaluation map e.*

In concrete terms the evaluation map can be given by a polynomial of total degree $6g-6$ in dual variables $\phi'_\lambda, \Omega', u'$. On symmetry grounds one can see that this is a polynomial in $\Lambda, \Omega', u'$, where $\Lambda \in R^*_6$ represents the intersection form of $\Sigma$; thus for large $g$ there are aproximately $3g^2/2$ coefficients to be determined.

In fact the work of Atiyah and Bott shows that $R$ maps onto the rational cohomology ring of $W$, so the latter ring is entirely determined by the ideal $I = \text{Ker } M$. In turn this ideal can be recovered in principle from the evaluation map $e$ and the multiplication in $R$, for by Poincaré duality we have:

$$M(\alpha) = 0 \iff e(\alpha\beta) = 0, \quad \text{for all } \beta \in R. \tag{12}$$

Conversely, if we know the ideal of relations $I$ we can recover the map $e$. Thus Problem 4 is equivalent to finding the rational cohomology ring of $W$, in terms of the canonical generators $\phi_\lambda, \Omega, u$. While there are many partial results in this direction, and the additive structure of the cohomology is completely known, this general problem has not been solved yet. (There is an extensive discussion of the problem in [2].) Another feature of the higher dimensional case discussed in Section 3 above which is already found in the 2-dimensional situation is the existence of a natural complex Kahler structure on the moduli space $W$. In this case the Kahler form represents the class $M(\omega)$ in $H^2(W)$. So the pairing $e(\Omega^{3g-3})$ is $(3g-3)!$ times the volume of the moduli space $W$. Again there is a determinant line bundle $\mathcal{L}$ over the moduli space with curvature proportional to the Kahler form. Here one makes contact with recent developments in conformal field theory, and in particular with Witten's interpretation of the Jones invariants [12]. In that theory one associates to the Riemann surface $\Sigma$ a series of complex vector spaces $H_k$, for positive integers $k$. These are the spaces of holomorphic cross sections of the line bundles $\mathcal{L}^{\otimes k}$ over $W$. Their dimensions can be found from the Riemann-Roch formula (since the higher cohomology groups are zero):

$$\begin{aligned} \dim H_k &= \chi(\mathcal{L}^{\otimes k}) \\ &= \langle \text{ch}(\mathcal{L}^{\otimes k})\text{Td}(W), [W]\rangle \\ &= k^{3g-3}\text{Vol}(W) + \text{O}(k^{3g-2}). \end{aligned} \tag{13}$$

Thus the pairing $e(\Omega^{3g-3})$ can be found from the leading term in $\dim H_k$, as a function of $k$, for large $k$. (The author has been told that these dimensions can be calculated explicitly, using the Verlinde fusion algebra.)

## 6. Construction of holomorphic bundles

We will now outline a standard algebro-geometric technique which can be used to construct holomorphic bundles over Riemann surfaces, and hence obtain information about moduli spaces. We begin by recalling that a rank 2 holomorphic bundle $V$ over $\Sigma$ is defined to be stable if any line subbundle $L$ of $V$ has degree $L \leq \frac{1}{2}$degree $V$ (the degree of

## 4. Instantons in Yang–Mills theory

a bundle over $\Sigma$ is its first Chern class, evaluated on the fundamental cycle). Now fix a line bundle $H$ over $\Sigma$ of degree 1. We will discuss stable rank 2 bundles $V$ with $\Lambda^2 V = H$ (which correspond to flat $SO(3)$ connections); for such a bundle the Riemann-Roch theorem gives that:

$$\chi(V \otimes H^p) = \dim H^0(V \otimes H^p) - \dim H^1(V \otimes H^p) = 2(p-g+1)+1, \quad (14)$$

so if $p \geq g - 1$ the bundle $V \otimes H^p$ has a non-trivial holomorphic section. If $s$ is such a section the zeros of $s$ define a divisor on $\Sigma$ and if we denote by $D$ the corresponding line bundle, $s$ gives an inclusion of $D \otimes H^{-p}$ as a subbundle of $V$. So for any such bundle $V$ we can find a line subbundle of degree less than or equal to $g - 1$. Now, switching notation, let $L$ be a line subbundle of $V$, so we have an exact sequence of bundles:

$$0 \longrightarrow L \longrightarrow V \longrightarrow H \otimes L^{-1} \longrightarrow 0. \quad (15)$$

This exact sequence defines an "extension class" in the first sheaf cohomology group $H^1(L^2 \otimes H^{-1})$. Conversely given a class $\beta$ in this cohomology group we can construct an exact sequence of the form (15), and hence in particular a holomorphic bundle $V_\beta$ as middle term. For any non-zero scalar $t$ the bundle $V_{t\beta}$ is isomorphic to $V_\beta$.

We can now explain the picture of the moduli space $W$ (of stable rank 2 bundles with fixed determinant) along the lines developed by Atiyah and Bott. We expect that generically (in the moduli space of stable bundles) the space of holomorphic sections $H^0(V \otimes H^p)$ will be zero for $p \leq g - 2$ and will be one dimensional for $p = g - 1$. More generally we expect that $V \otimes U$ will have no non-trivial sections for any line bundle $U$ of degree less than $g - 1$. Such bundles then are associated to a unique point in the projective space $\mathbf{P} = \mathbf{P}(H^1(\Sigma, H^{-(1+2g)}))$. Conversely, for generic points $\beta$ in $\mathbf{P}$ the corresponding bundle $V_\beta$ will be stable. Thus we can identify a dense open set in $\mathbf{P}$ with a corresponding open set in $W$. More precisely we get a birational isomorphism between $W$ and $\mathbf{P}$. We can think of the $(3g - 3)$-dimensional projective space $\mathbf{P}$ as being a "first approximation" to the moduli space $W$. To construct the actual moduli space from this point of view we have to perform a sequence of modifications, analogous to "surgeries" in manifold topology. We have to cut out from $\mathbf{P}$ the points $\beta$ for which $V_\beta$ is unstable and glue in points corresponding to bundles constructed from extensions (15) with degree $L \leq g - 2$. Also we must collapse some subvarieties in $\mathbf{P}$, corresponding to bundles $V$ with two or more independent sections of $V \otimes H^{g-1}$. There is also, as explained by Atiyah and Bott, a description of this approximation procedure from the point of view of connections and Yang-Mills theory. The projective space **P** can be regarded as a submanifold of $\mathcal{A}/\mathcal{G}$ consisting of a family of connections close to the reducible connection on $H^{1-g} \oplus H^g$. For these the Yang-Mills action (suitably normalised) is approximately $g^2 + (1-g)^2 - 1$. We now consider the gradient flow of the Yang-Mills action:

$$\frac{\partial A}{\partial t} = -d_A^* F_A. \qquad (16)$$

Under this evolution most points of **P** flow down to the absolute minimum of the action, the manifold $W$, viewed now as parametrising projectively flat connections with normalised action 0. This process gives the same identification between open sets in **P** and $W$ as the algebro-geometric construction above. The complications come from the "exceptional" points of **P** which do not flow down to the absolute minimum, any such point flows down to a reducible connection associated to a direct sum $L^{-1} \oplus L \otimes H$ with degree $L = p < g - 1$, and normalised action $p^2 + (1-p)^2 - 1$. These correspond to the unstable bundles which are extensions of $L^{-1}$ by $L \otimes H$.

## 7. Bundles over algebraic surfaces

We now return to the case of primary interest in this paper, moduli spaces of stable $SL(2, \mathbf{C})$ bundles over a complex algebraic surface $S$, which correspond to $SU(2)$ instantons. We assume that the surface $S$ is simply connected and that the Picard group of line bundles over $S$ has rank 1, the latter is commonly the case for simply connected surfaces. So the line bundles over $S$ all have the form $H^p$, where $H$ is a positive line bundle over $S$. We write $h$ for $c_1(H)$ and $\kappa$ for the canonical class $c_1(K_S)$. There is a variant of the procedure discussed in Section 6 above for constructing bundles over complex surfaces. Given any bundle $V$ we can again find a minimal $p$ such that $V \otimes H^{-p}$ has a non-trivial holomorphic section $s$. Thus $s$ vanishes on a set $Z$ of isolated points in $S$. The new feature is that we cannot now twist to get a subbundle of $V$. Instead our section exhibits $V$ as an extension:

$$0 \longrightarrow H^{-p} \longrightarrow V \longrightarrow H^p \otimes I \longrightarrow 0, \qquad (17)$$

where $I$ is the ideal sheaf defined by the zeros $Z$. Going in the other direction, for fixed $p$ and $Z$ we can construct an extension, and hence a rank 2 bundle as middle term, from a suitable class $\beta$ in a vector space :

$$\mathbf{E} = \mathbf{Ext}_1(H^p \otimes I, H^{-p}). \qquad (18)$$

## 4. Instantons in Yang–Mills theory

This theory is fully explained in [8, Chapter 5]. Roughly speaking, **E** consists of 1-dimensional cohomology classes over $S - Z$ with certain singularities (analogous to poles) at the points of $Z$. The singularity at a point $z$ in $Z$ is determined by a residue $r$ which lies in a copy of the complex numbers. More precisely $r$ lies in the fibre over $z$ of the line bundle $(H^{2p} \otimes K_S)^{-1}$. To obtain a bundle as the middle term of the exact sequence we require $r$ to be non-zero. Again, multiplication by scalars gives equivalent bundles so, with $p$ and $k = c_2(V)$ fixed we have $3l - 1$ complex parameters determining the singularity data where :

$$l = \text{No. of points in } Z = c_2(V \otimes H^p) = k + h^2p^2. \qquad (19)$$

That is, we have two parameters determining each point of $Z$ in $S$ and one parameter specifying the residue at that point, giving $3l$, then we subtract one for the overall action of the scalars. We denote the points of $Z$ by $\{z_1, \ldots, z_l\}$ and the residues by $\{r_1, \ldots r_l\}$. It is easy to show that, for large $p$, a class in **E** is uniquely determined by the singularity data. However this data is subject to constraints imposed by the holomorphic sections of the line bundle $H^{2p} \otimes K_S$. For any section $\sigma$ of this line bundle we can pair $r_i$ with the value of $\sigma$ at $z_i$ to get a complex number $\langle r_i, \sigma \rangle$. The constraints on the $r_i$ are that for every such section we have:

$$\sum_{i=1}^{l} \langle r_i, \sigma \rangle = 0. \qquad (20)$$

(This equation falls under the general slogan "the sum of residues is zero", and can be proved using Stokes' theorem.) Using this theory we can in principle, construct all stable bundles and so the moduli spaces $M_k$, provided we know enough about the sections of the line bundles over $S$. The Riemann-Roch formula is now:

$$\chi(V \otimes H^p) = p^2 h^2 - ph.\kappa + 2\chi - k, \qquad (21)$$

where $\chi = \chi(S) = 1 + p_g(S)$ is the holomorphic Euler characteristic of $S$. We cannot now always choose $p$ so that the Euler characteristic $\chi(V \otimes H^p)$ is 1, but for some values of $k$ this will be possible, i.e. when

$$k = p^2 h^2 - ph.\kappa + 2\chi - 1. \qquad (22)$$

In this case it is reasonable to expect that for a generic stable bundle $V$ there will be a 1-dimensional space of sections of $V \otimes H^p$ and that these sections will vanish on a set of isolated points $Z$ with :

$$l = \text{No. of points in } Z = 2p^2 h^2 - ph.\kappa + 2\chi - 1. \qquad (23)$$

We then get a birational equivalence between the moduli space $M_k$ and a space $N_k$ which parametrises equivalence classes of triples $(V, \phi, s)$, where $V$ is a holomorphic bundle, $\phi$ is a trivialisation of $\Lambda^2(V)$, $s$ is a section of $V \otimes H^p$ having isolated zeros, and we identify $(V, \phi, s)$ with $(V, \phi, ts)$, for $t$ in $\mathbf{C}^*$. Then $N_k$ can be identified with systems of singularity data $(r_i, z_i)$, satisfying the constraints, as described above.

## 8. Topological calculations

To what extent can this description be used to calculate the differential topological invariants defined by the moduli spaces? It is a truism in algebraic geometry that it is easier to calculate the number of solutions to a system of equations, counted with appropriate multiplicities where necessary, than to find the solutions explicitly — the simplest example being the solutions of polynomial equations in one complex variable. More generally, it is easier to calculate the homology class of a solution set, in some ambient space, than to describe the set explicitly. In this spirit one can hope that it should be possible to calculate homological invariants of the moduli spaces, without having a detailed description of them. To this end we begin by considering, in the situation of Section 7 above, the moduli space $N_k$.

First one should construct a space $B = B(p)$ which parametrises all choices of singularity data, that is, roughly speaking, sets of $l$ points $\{z_i\}$ and residues $r_i$ at these points (with $l$ given by (23)). This is straightforward if we work with sets of distinct points. We take the vector bundle $\bigoplus_{i=1}^l \pi_i^*((H^{2p} \otimes K_S)^{-1})$ over the product of $l$ copies of $S$, minus the "diagonals" (here $\pi_i$ denotes the projection onto the $i$th factor). We divide by the natural action of the permutation group to get a bundle $F$ over the configuration space $C^l(S)$ of $l$ distinct points in $S$. Now take the associated projective bundle $\mathbf{P}(F)$. This is a dense open subset of $B$, to obtain the complete space one needs to understand multiple points and introduce appropriate sheaves. We assume this can be done, in such a way that $B$ is a compact complex variety of dimension $3l - 1$. Over $B$ there is a line bundle $U$ extending the hyperplane bundle over $\mathbf{P}(W)$.

The moduli space $N_k$ is now naturally embedded as a complex subvariety of $B$. It is defined by two conditions, first we have the closed constraints imposed by the sections of $H^{2p} \otimes K_S$ over $S$. Each such section $\sigma$ induces a section $\tau_\sigma$ of $U$ over $B$, and the moduli space $N_k$ is contained in the common zero set, $\overline{N}_k$ say, of the $\tau_\sigma$. Note that the dimensions behave in the correct way: the number of independent sections of $H^{2p} \otimes K_S$ is, for large $p$, given by Riemann-Roch:

$$\dim H^0(H^{2p} \otimes K_S) = \chi(H^{2p} \otimes K_S) = 2p^2 + ph.\kappa + \chi = n, \text{ say.} \quad (24)$$

## 4. Instantons in Yang–Mills theory

So we expect the dimension of $\overline{N}_k$ to be:

$$\begin{aligned}\dim \overline{N}_k &= \dim B - n = 3l - n - 1 \\ &= 4p^2h^2 - 4ph.\kappa + 5\chi - 4 = 4k - 3\chi,\end{aligned} \qquad (25)$$

and $4k - 3\chi$ is indeed the (complex) dimension $d$ of the moduli space $M_k$ for large $k$. Second, we have to take into account open non-degeneracy conditions. Again this is straightforward in the open subset $\mathbf{P}(W)$ of $B$—we remove the points of $\overline{N}_k$ which lie in the "co-ordinate hyperplanes" in the projective space fibres, corresponding to the fact that each of the residues $r_i$ must be non-zero. Thus $N_k$ is an open subset of the compact space $\overline{N}_k$, which is a complete intersection cut out by $n$ sections of the line bundle $U$ over $B$.

The universal cohomology classes $\mu(\alpha)$ are defined over any family of bundles on $S$, and so in particular over $N_k$. For a 2-dimensional homology class $\alpha$ on $S$ the corresponding cohomology class $\mu(\alpha)$ over $N_k$ can be identified as follows. There is a projection map $\varpi : B \to s^l(S)$, to the $l$-fold symmetric product of $S$. Over $s^l(S)$ we have a "symmetric sum" $s_l(\alpha)$ in $H^2(s^l(S))$. Then $\mu(\alpha)$ is the restriction of the class:

$$\mu'(\alpha) = \langle h, \alpha \rangle c_1(U) + \varpi^*(s_l(\alpha)) \qquad (26)$$

on $B$ to $N_k$. Now the space $\overline{N}_k$, as the complete intersection cut out by $n$ sections of $U$, is Poincaré dual to $c_1(U)^n$. So we are led to propose the formula:

$$\langle \mu(\alpha)^d, [\overline{N}_k] \rangle = \langle u^n(\langle h, \alpha \rangle u + \varpi^*(s_l(\alpha)))^d, [B] \rangle. \qquad (27)$$

Here we have written $u$ for $c_1(U) \in H^2(B)$. The point of this formula is that the right hand side depends only on the structure of the space $B$ and the class $u$ in $H^2(B)$, whereas the left hand side is clearly related to the pairing $\langle \mu(\alpha)^d, [M_k] \rangle$ which we want to calculate.

What is the relation between the pairings of the class $\mu(\alpha)^d$ with $M_k$ and $N_k$? On the one hand $M$ and $N$ are birationally equivalent, so have a common dense open subset. On the other hand we know that in the similar situation of bundles over a Riemann surface the corresponding birational approximation to the moduli space is a rather poor model from the point of view of the homology pairings. In either situation, to get the true moduli space we have to perform "surgeries" in which we remove the unstable bundles and add in bundles constructed from a lower twisting. However in the case at hand a general position argument, based on counting the expected dimensions of the parts of the moduli space constructed from smaller twists, indicates that for any class $\alpha$ with $\langle h, \alpha \rangle = 0$ the pairings of $\mu(\alpha)^d$ with $M_k$ and the completion $\overline{N}_k$ of $N_k$ agree. So, tentatively, we propose that (27) gives a formula for the invariants $\gamma(\alpha)$ on classes $\alpha$ annihilated by $h$.

## 9. A problem in enumerative geometry

We now go on to make the conclusions of the discussion above (which, we should emphasise, have certainly not been established rigorously) more explicit. Consider first the case when $S$ is a K3 surface, so the canonical bundle $K_S$ is trivial and $\kappa = 0$. We suppose that the Picard group is generated by a positive line bundle $H$ with $c_1(H)^2 = h^2 = 2$. Then we consider rank-2 bundles $V$ with $c_2 = k$ where:

$$k = 2p^2 + 3 \tag{28}$$

for large integers $p$. Then we expect that generically there will be a one-dimensional space of sections of $V \otimes H^p$. Now our formulae (23), (24) give in this case that the dimension $n$ of the constraint space is $4p^2 + 2$, while the number $l$ of zeros of a section is $4p^2 + 3$. Thus $n = l - 1$ is the dimension of the general fibre of the map $\varpi$ from $B$ to $s^l(S)$. So for a general choice of $l$ points in $S$ we expect there to be a unique set of residues satisfying the constraints. That is, we expect the projection map $\varpi$ to give a birational isomorphism from $N$ to the symmetric product $s^l(S)$. (Note that the complex dimension $d$ of the moduli space is equal to $2l$.) Thus our pairing $\langle \mu(\alpha)^d, [N] \rangle$ should be equal to $\langle (s_l(\alpha))^d, [s^l(S)] \rangle$, for classes $\alpha$ annihilated by $h$. We evaluate this latter pairing by a combinatorial argument: let $\Sigma$ be a representative in $S$ for the homology class $\alpha$, the class $s_l(\alpha)$ on the symmetric product is represented by the set $s_l(\Sigma)$ of configurations $(x_i)$ with at least one of the points $x_i$ lying on $\Sigma$. Now choose a set of $d = 2l$ representatives $\Sigma_\lambda$ in general position; the evaluation of the cup product is given by the intersection of all the corresponding subsets $s_l(\Sigma_\lambda)$ in the symmetric product, that is we must count (with appropriate signs) the number of ways of placing $l$ points in $S$ such that each surface $\Sigma_\lambda$ contains a point. The possible configurations are just those where each point lies on an intersection of two of the surfaces:

$$x_1 \in \Sigma_{\lambda_1} \cap \Sigma_{\lambda_2}, \quad x_2 \in \Sigma_{\lambda_3} \cap \Sigma_{\lambda_4}, \ldots \tag{29}$$

for a permutation $\lambda_1 \lambda_2 \ldots \lambda_{2l}$ of $1\,2\ldots 2l$. Counting with signs we get the evaluation:

$$\langle (s_l(\alpha))^d, [s^l(S)] \rangle = \frac{(2l)!}{2^l l!} (\alpha.\alpha)^l. \tag{30}$$

Our tentative general discussion thus indicates that, on the classes annihilated by $h$, the polynomial invariant $\gamma(\alpha)$ defined by the moduli space $M_k$ is just $\frac{(2l)!}{2^l l!}$ times the $l$th power of the intersection form of $S$, where $l = 2k - 3$. In fact on symmetry grounds one knows that the polynomial invariants for $S$ are always multiples of powers

## 4. Instantons in Yang–Mills theory

of the intersection form, so we expect the invariant to be given by this formula on all classes, not just those annihilated by $h$. Now, fortunately, this agrees with a rigorous calculation of the invariants of a K3 surface obtained recently by Friedman and Morgan [7], using another approach which exploits the fact that there are "elliptic" K3 surfaces. Returning to the discussion of the first part of the paper we obtain, modulo the solution to Problem 3, the fact that for a K3 surface $S$, with a Kahler metric, the volume of the moduli space $M_k$ is:

$$\mathrm{Vol}(M_k) = \frac{1}{l!}\left(\mathrm{Vol}(S)\right)^l, \tag{31}$$

where $l = 2k - 3$.

We now turn to consider a surface $S$ of "general type" with one-dimensional Picard group, generated by a positive fractional multiple $H$ of the canonical bundle. We have then

$$n - (l-1) = 2p(h.\kappa) - (\chi - 2). \tag{32}$$

So, for large $p$, the space $N$ has a lower dimension than the symmetric product. We can express the calculation of the pairing (27) in the same combinatorial fashion as above: choosing $d$ representatives $\Sigma_\lambda$ we want to count, with signs, the configurations of $l$ points $(x_i)$ which lie on an $n-l$ dimensional family of curves in the linear system $|2pH + K_S|$ (with $n,l$ given by (23), (24)) and for which each surface $\Sigma_\lambda$ contains one of the points $x_i$. There are a number of classes of possible configurations to consider but it seems that, if $\alpha$ is annihilated by $h$, the only ones which give a contribution to the algebraic sum are those configurations where $d/2$ of the points lie on intersections of the $\Sigma_\lambda$ and the remainder are constrained only by the linear system $|2pH + K_S|$ (here $d = 3l - 1 - n$, and we suppose $d$ is even). We can then cast our problem into the following geometric form. For a given set of intersection points $(x_i)_{i=1}^{d/2}$ we consider the linear system of sections of $2pH + K_S$ vanishing at all the $x_i$, $i = 1,\ldots d/2$. This linear system gives a rational map from $S$ to $\mathbf{P}^r$ where $r = n - d/2 - 1$. The image $T$ has degree $(2ph + \kappa)^2$. Then the number of ways of completing the set $x_1,\ldots x_{(d/2)}$ to a configuration of the desired kind, by points $x_{(d/2)+1}\ldots x_l$, is the number of configurations of $m = l - d/2$ points on $T$ which lie on a linear subspace of $\mathbf{P}^r$ of dimension $m - 2$. Notice that $r = 3m - 2$.

We can now sum up our discussion by the following conjectures. For any $m \geq 1$ and a surface $T$ in projective space of dimension $3m-2$ let $\nu(T)$ be the number of configurations of $m$ points in $T$ which lie in a linear subspace of dimension $m-2$, the configurations being counted with suitable multiplicities if necessary. (We assume $T$ is sufficiently generic for there to be only a finite number of such configurations.)

**Conjecture 5** *There is a universal formula expressing $\nu(T)$ in terms of $m$, the Chern numbers of $T$, the degree of $T$ in $\mathbf{P}^{3m+2}$, and the intersection number of the canonical class of $T$ with the restriction of the hyperplane class.*

**Conjecture 6** *Let $S$ be a surface whose Picard group is generated by a positive line bundle $H$ and write $h = c_1(H)$, $\kappa = c_1(K_S)$. Then for large integers $p$, and $k$ equal to $p^2 h^2 - ph.\kappa + 2\chi(S) - 1$ the polynomial function $\gamma(\alpha)$ defined by the moduli space $M_k$ of $SU(2)$ instantons over $S$ with $c_2 = k$ has the form $\gamma(\alpha) = \gamma_1(\alpha) + \gamma_2(\alpha) h(\alpha)$, where $\gamma_1(\alpha)$ is $0$ if $\chi(S)$ is odd and otherwise is equal to $\frac{(2s)!}{2^s s!} \nu(T)(\alpha.\alpha)^s$, for $s = 2k - \frac{3}{2}\chi(S)$, and $T$ is a surface of degree $(2ph + \kappa)^2$ in $\mathbf{P}^r$, with $r = 3(ph.\kappa - \chi(S) + 1) + 2$, biholomorphic to $S$ blown up at $s$ points and embedded by the linear system $|2pH + K_S|$.*

If the arguments we have sketched stand up to a more rigorous treatment the calculation of these components of the Yang-Mills invariants will thus be equivalent to the problem in enumerative geometry of evaluating the integers $\nu(T)$ for surfaces in projective space, this latter problem is interesting in its own right and has not, so far as the author knows, been treated in the algebraic geometry literature.

**References**

[1] M.F. Atiyah, On Donaldson-Witten Theory, private communication, 1988.

[2] M.F. Atiyah and R. Bott, The Yang-Mills equations over Riemann surfaces, *Phil. Trans. Roy. Soc. London* (Series A) **308** (1982) 523–615.

[3] S.K. Donaldson, Connections, cohomology and the intersection forms of 4-manifolds, *J. Diff. Geom.* **24** (1986) 275–341.

[4] S.K. Donaldson, Infinite determinants, stable bundles and curvature, *Duke Math. Jour.* **54(1)** (1987) 231-247.

[5] S.K. Donaldson, Polynomial invariants for smooth 4-manifolds, *Topology*, to appear.

[6] S.K. Donaldson, Compactification and completion of Yang–Mills moduli spaces, in *Proceedings International Symposium on Differential Geometry*, Peñiscola, 1988, Springer Lecture Notes in Mathematics, to appear.

[7] R. Friedman and J. Morgan, Complex versus Differentiable classification of algebraic surfaces, *Columbia University Preprint*, 1988.

[8] P. Griffiths and J. Harris, *Principles of Algebraic Geometry*, John Wiley, New York, 1978.

[9] D. Groisser and T. Parker, The Riemannian Geometry of the Yang–Mills moduli spaces, *Commun. Math. Physics* **112** (1987) 663–689.

[10] V. Mathai and D. Quillen, Superconnections, Thom classes, and the Chern character, *Topology* **25** (1986)85–110.

[11] E. Witten, Topological Quantum Field Theory, *Commun. Math. Phys.* **117** (1988)353–396.

[12] E. Witten, Some geometrical applications of quantum field theory, in *Proceedings of IXth IAMP Congress, Swansea 1988*, ed. B. Simon, A. Truman and I.M. Davies, Adam Hilger, Bristol and New York, 1989.

MATHEMATICAL INSTITUTE, 24-29 ST.GILES', OXFORD OX1 3LB, ENGLAND.

# 5

## Extended Conformal Algebras

*Peter Goddard*

### 1. Introduction

The algebra of the group of conformal transformations in two dimensions consists of two commuting copies of the *Virasoro algebra* [1], $\hat{v}$. It has a basis consisting of $L_n$, $n \in \mathbf{Z}$, and a central element, $c$, satisfying the commutation relations,

$$\begin{aligned} [L_m, L_n] &= L_{m+n} + \tfrac{c}{12}m(m^2 - 1)\delta_{m,-n}, \\ [L_n, c] &= 0 \end{aligned} \quad (1)$$

In many mathematical and physical contexts, the representations of $\hat{v}$ which are relevant satisfy two conditions: they are *unitary*, i.e. they satisfy the hermiticity condition

$$L_n^\dagger = L_{-n}, \quad (2)$$

with respect to a positive definite scalar product; and they have the *"positive energy"* property that $L_0$ is bounded below. In an irreducible unitary representation the central element $c$ takes a fixed real value.

In physical contexts, the value of $c$ is a characteristic of a theory. If $c < 1$, it turns out that the conformal algebra (1) is sufficient to "solve" the theory, in the sense of relating the calculation of the infinite set of physically interesting quantities to a finite subset which can be handled in principle. For $c \geq 1$, this is no longer the case for the algebra (1) alone and one needs some sort of extended conformal algebra, such as the superconformal algebra. It is these algebras that this paper aims at addressing.

### 2. Conditions for unitarity

We shall be interested in positive energy representations of the Virasoro algebra (1), *i.e.* those for which $L_0$ is diagonalisable and its spectrum is bounded below. This is a typical physical requirement because $L_0$ often corresponds to an energy or dilatation operator, or something similar. Then, since

$$L_0 L_n = L_n(L_0 - n), \quad (3)$$

the $L_n$, $n > 0$, act as lowering operators for the eigenvalues of $L_0$. Thus, if $|h\rangle$ is an eigenvector of $L_0$ corresponding to the lowest eigenvalue $h$,

$$L_0|h\rangle = h|h\rangle, \\ L_n|h\rangle = 0, \quad n > 0. \qquad (4)$$

Such a state $|h\rangle$ is usually called a *highest weight state*. It is not difficult to see that the whole of an irreducible positive energy representation is built up from such a highest weight state by the action of the algebra, and that the representation space is spanned by states of the form

$$(L_{-1})^{n_1}(L_{-2})^{n_2}\ldots(L_{-r})^{n_r}|h\rangle. \qquad (5)$$

Such a representation is called an irreducible *highest weight representation* and it is characterised by the pair of numbers $(c, h)$. Any unitary positive energy representation is the direct sum of highest weight representations.

The necessary [2] and sufficient [3] condition for a highest weight representation to be unitary is that *either*

$$c \geq 1, \quad h \geq 0, \qquad (6)$$

or

$$c = 1 - \frac{6}{(m+2)(m+3)}, \quad m = 0, 1, \ldots, \qquad (7)$$

$$h = \frac{[(m+3)p - (m+2)q]^2 - 1}{4(m+2)(m+3)}, \qquad (8)$$

where $p = 1, 2, \ldots m+1$ and $q = 1, 2, \ldots p$.

## 3. Modular Invariance

The representations with $c < 1$ are particularly interesting. We define the character, $\chi_{c,h}(\tau)$, of the representation $(c, h)$ of the Virasoro algebra by

$$\chi_{c,h}(\tau) = q^{-\frac{c}{24}} \text{ trace}\{q^{L_0}\}. \qquad (9)$$

where

$$q = e^{2\pi i \tau}. \qquad (10)$$

For a given value of $c < 1$, the $\chi_{c,h}(\tau)$ for the various values of $h$ transform into linear combinations of one another under the modular transformations $\tau \mapsto -1/\tau$ and $\tau \mapsto \tau + 1$. [These generate a discrete group, the *modular group*,

$$\tau \mapsto \tau' = \frac{a\tau + b}{c\tau + d}, \qquad (11)$$

## 5. Extended Conformal Algebras

where $a, b, c$ and $d$ are integers and $ad - bc = 1$, which maps the upper half complex $\tau$-plane into itself.] In fact the $\chi_{c,h}(\tau)$, for given $c < 1$, can be regarded as the basis functions for finite-dimensional unitary representation of the modular group. Thus

$$\sum_h \chi_{c,h}(\tau)^* \chi_{c,h}(\tau) \tag{12}$$

is a modular invariant. (The asterisk denotes complex conjugation.)

For $c > 1$,

$$\chi_{c,h}(\tau) = q^{-\frac{c}{24}+h} \prod_{n=1}^{\infty} (1 - q^n)^{-1}, \tag{13}$$

and it can be shown that an infinite number of $h$ values have to be combined to get a modular invariant. (12) is the form that the partition function takes a "non-chiral theory"; more generally it can have the form

$$\sum_{h,\bar{h}} N_{h\bar{h}} \chi_{c,\bar{h}}(\tau)^* \chi_{c,h}(\tau), \tag{14}$$

where the $N_{h\bar{h}}$ are non-negative integers with $N_{00} = 1$ and $h$ and $\bar{h}$ each range over the values permitted by (14) for the given value of $c < 1$. Cardy [4] has shown that a modular invariant theory will be finitely reducible with respect to the conformal group if and only if $c < 1$. In conformal field theory terms, this means that a modular invariant theory has a finite number of primary fields with respect to the conformal group if and only if $c < 1$. What should one do if $c \geq 1$? A natural approach is to try to extend the conformal symmetry algebra into one which is finitely reducible for the given value of $c$.

### 4. The super-Virasoro algebra

A natural extension to consider is the *super-Virasoro algebra*,

$$[L_m, L_n] = (m-n)L_{m+n} + \tfrac{c}{12}m(m^2-1)\delta_{m,-n}$$

$$[L_m, G_r] = (\tfrac{m}{2} - r)G_{m+r} \tag{15}$$

$$\{G_r, G_s\} = 2L_{r+s} + \tfrac{c}{3}(r^2 - \tfrac{1}{4})\delta_{r,-s}$$

where $m, n \in \mathbf{Z}$ and either $r, s \in \mathbf{Z} + \tfrac{1}{2}$ (NS case) or $r, s \in \mathbf{Z}$ (R case). Again unitary positive energy representations are labelled by $(c, h)$ and we have similar restrictions to those in (6) (7) (8), *either*

$$c \geq \frac{3}{2}, \qquad h \geq 0, \tag{16}$$

*or*

$$c = \frac{3}{2}\left(1 - \frac{8}{(m+2)(m+4)}\right), \tag{17}$$

with $h$ restricted to a finite set of values for each $c$ in this discrete series.

A superconformal theory will be finitely reducible with respect to the super-Virasoro algebra if $c < \frac{3}{2}$; what is available if $c \geq \frac{3}{2}$? Amongst the possibilities are the parafermionic algebras introduced by Fateev and Zamolodchikov [5].

## 5. Parafermion algebras

Parafermionic algebras are not defined by commutators or anticommutators, but by operator product equations. The Virasoro algebra itself is equivalent to assuming the operator product expansion

$$L(z)L(\zeta) = \frac{\frac{1}{2}c}{(z-\zeta)^4} + \frac{2L(\zeta)}{(z-\zeta)^2} + \frac{L'(\zeta)}{(z-\zeta)} + \mathcal{O}(1), \qquad (18)$$

for $|z| > |\zeta|$, where the prime denotes differentiation and

$$L(z) = \sum_n L_n z^{-n-2}. \qquad (19)$$

An example of a parafermionic algebra, introduced by Fateev and Zamolodchikov, is the $S_3$ algebra specified by (18) together with the operator product expansions:

$$\begin{aligned}
A(z)A(\zeta) &= \lambda(z-\zeta)^{-\frac{4}{3}}\left\{\bar{A}(\zeta) + \frac{1}{2}(z-\zeta)\bar{A}'(\zeta) + \mathcal{O}\left((z-\zeta)^2\right)\right\} \\
A(z)\bar{A}(\zeta) &= (z-\zeta)^{-\frac{8}{3}}\left\{1 + \frac{8}{3c}(z-\zeta)^2 L(\zeta) + \mathcal{O}\left((z-\zeta)^3\right)\right\} \quad (20)\\
L(z)A(\zeta) &= (z-\zeta)^{-2}\left\{\frac{4}{3}A(\zeta) + (z-\zeta)A'(\zeta) + \mathcal{O}\left((z-\zeta)^2\right)\right\}
\end{aligned}$$

together with similar expansions with the roles of $A(z)$ and $\bar{A}(z)$ are interchanged. In (20),

$$\lambda = \frac{2}{3}\sqrt{\frac{8-c}{c}}. \qquad (21)$$

Fateev and Zamolodchikov suggested parameterising $c$ in the form

$$c = 2\left(1 - \frac{3}{(k+1)(k+3)}\right), \qquad (22)$$

and that there might be unitary representations for $k$ a positive integer, or perhaps for $2k$ a positive integer. It can be shown that the weaker condition is sufficient using the "coset construction", which we shall move towards discussing.

## 6. The Sugawara construction

We can construct representations of the Virasoro algebra, and eventually of extensions of it, starting with representations of affine Kac-Moody algebras. Given a compact Lie algebra $g$, with a basis consisting of $t^a$, $1 \le a \le \dim g$, satisfying the commutation relations

$$[t^a, t^b] = i f^{abc} t^c, \tag{23}$$

where the structure constants $f^{abc}$ are totally antisymmetric, the affine Kac-Moody algebra $\hat{g}$ associated with $g$ is

$$[T_m^a, T_n^b] = i f^{abc} T_{m+n}^c + k\, m\, \delta_{m,-n}\, \delta^{ab}, \tag{24}$$

where $k$ is a central element,

$$[T_m^a, k] = 0. \tag{25}$$

A representation of $\hat{g}$ can be extended to a representation of the semi-direct product of $\hat{v}$ and $\hat{g}$ is demonstrated by the Sugawara [6] construction which represents the Virasoro algebra as bilinear quantities in the Kac-Moody generators, roughly

$$\Theta(z) \equiv \sum_{n \in \mathbf{Z}} L_n z^{-n-2} \sim T^a(z) T^a(z), \tag{26}$$

where a sum over $a$ is implied and

$$T^a(z) = \sum_n T_n^a z^{-n-1}. \tag{27}$$

To make this precise, we need to be careful about such matters as normal ordering, so that the expressions are well-defined. We define a normal ordering operation by

$$\begin{aligned} {}^{\times}_{\times} T_m^a T_n^a {}^{\times}_{\times} &= T_m^a T_n^a && \text{if } n \ge 0, \\ &= T_n^a T_m^a && \text{if } n \le 0. \end{aligned} \tag{28}$$

Note that this definition is phrased entirely in terms of the operators $T_n^a$ and is not in any way dependent on the way they might be constructed out of any more basic oscillators. First, as a precise interpretation of (26), consider

$$\tilde{\mathcal{L}}_n^g = \tfrac{1}{2} \sum_m {}^{\times}_{\times} T_m^a T_{n-m}^a {}^{\times}_{\times}. \tag{29}$$

Calculation shows that this is not quite a representation of the Virasoro algebra. First, from (24), one can deduce that

$$[\tilde{\mathcal{L}}_m^g, T_n^a] = -(k + \tfrac{1}{2} Q^g) n T_{m+n}^a, \tag{30}$$

where $Q^g$ is the quadratic Casimir operator of $g$ in the adjoint representation,
$$f^{abc}f^{abd} = Q^g \delta^{cd}. \tag{31}$$
Thus we are led to renormalise the definition (29) to give
$$\mathcal{L}_n^g = \frac{1}{2k + Q^g} \sum_m {}^\times_\times T_m^a T_{n-m}^a {}^\times_\times. \tag{32}$$
Then, it follows from (30) that
$$[\mathcal{L}_m^g, T_n^a] = -nT_{m+n}^a, \tag{33}$$
and
$$[\mathcal{L}_m^g, \mathcal{L}_n^g] = (m-n)\mathcal{L}_{m+n}^g + \frac{c^g}{12}m(m^2-1)\delta_{m,-n}, \tag{34}$$
where
$$\begin{aligned} c^g &= \frac{2k \dim g}{2k + Q^g} \\ &= \frac{x \dim g}{x + h^g}. \end{aligned} \tag{35}$$
We have used
$$h^g = \frac{Q^g}{\psi^2}, \tag{36}$$
where $\psi$ denotes a long root of $g$, to denote the *dual Coxeter number* of $g$, which is always an integer, and
$$x = \frac{2k}{\psi^2} \tag{37}$$
is called the *level* of the representation of $\hat{g}$, and is also always an integer for unitary highest weight representations.

From (35) it is clear that the value of the central charge $c^g$ provided by the Sugawara construction is always rational but it is not in general integral or even half-integral as it would be for free bosons or free fermions. In fact, it is not difficult to see that
$$\mathrm{rank} g \leq c^g \leq \dim g. \tag{38}$$
If $g$ is an abelian algebra, these inequalities become equalities.

## 7. The Coset Construction

Since the Sugawara construction always gives a value of $c$ bigger than 1, to construct the representations in the discrete series (7), (8) we must use an elaboration [3] of the Sugawara construction. This construction is based on an algebra and a subalgebra, $g \supset h$, and so can

## 5. Extended Conformal Algebras

be associated with a coset space $g/h$. Given such a pair, we can use the Sugawara construction to define two Virasoro algebras, $\mathcal{L}_n^g$ and $\mathcal{L}_n^h$, with corresponding values of $c$, denoted $c^g$ and $c^h$, given by the general formula (35). The level $y$ of $\hat{h}$ is determined by the level $x$ of $\hat{g}$ together with the index $I$ describing the embedding $h \subset g$, $y = Ix$. It then follows that if $T_n^a$ is an element of $\hat{h}$, the relation (33) will hold in respect of both Virasoro algebras, i.e.

$$\begin{aligned}[] [\mathcal{L}_m^g, T_n^a] &= -nT_{m+n}^a, \\ [\mathcal{L}_m^h, T_n^a] &= -nT_{m+n}^a, \end{aligned} \qquad (39)$$

but this will not hold if $T_n^a$ is a member of $\hat{g}$ that is not in $\hat{h}$. From (39) it immediately follows that $\mathcal{L}_m^g - \mathcal{L}_m^h$ commutes with the whole of $\hat{h}$. We define

$$\mathcal{K}_n = \mathcal{L}_n^g - \mathcal{L}_n^h. \qquad (40)$$

Since $\mathcal{L}_m^h$ is constructed out of the elements of $\hat{h}$, it follows that it commutes with $\mathcal{K}_n$. Then, since

$$\begin{aligned} \mathcal{L}_n^g &= \mathcal{L}_n^h + \mathcal{K}_n, \\ [\mathcal{L}_m^g, \mathcal{L}_n^g] &= [\mathcal{L}_m^h, \mathcal{L}_n^h] + [\mathcal{K}_m, \mathcal{K}_n]. \end{aligned}$$

From this it is easy to see that $\mathcal{K}_m$ satisfies the Virasoro algebra,

$$[\mathcal{K}_m, \mathcal{K}_n] = (m-n)\mathcal{K}_{m+n} + \frac{c^{\mathcal{K}}}{12}m(m^2-1)\delta_{m,-n}, \qquad (41)$$

where the value of the central term is given by

$$c^{\mathcal{K}} = c^g - c^h. \qquad (42)$$

Note that (41) expresses $\mathcal{L}_n^g$ as the sum of two commuting Virasoro algebras.

In the above we have implicitly assumed that both $g$ and $h$ are simple algebras. However in practice we wish to apply the construction in cases in which either or both are semi-simple. To extend it to cover these cases we only need to describe how the Sugawara construction applies in semi-simple cases. This is quite straightforward. If $g$ is the direct sum of simple algebras $g_j$,

$$g = g_1 \oplus g_2 \oplus \ldots \oplus g_M, \qquad (43)$$

we apply the Sugawara construction to each of the simple components and add the results

$$\mathcal{L}_n^g = \mathcal{L}_n^{g_1} + \mathcal{L}_n^{g_2} + \ldots + \mathcal{L}_n^{g_M}. \qquad (44)$$

This provides a Virasoro algebra with a value of $c$ given by

$$c^g = c^{g_1} + c^{g_2} + \ldots + c^{g_M}, \tag{45}$$

where each $c^{g_j}$ is given by (35)

$$c^{g_j} = \frac{x_j \dim g_j}{x_j + h^{g_j}}, \tag{46}$$

where $h^{g_j}$ is the dual Coxeter number of $g_j$ and $x_j$ is the level of the representation of $\hat{g}_j$. (To specify a highest weight unitary representation of $g$ we need to specify the lowest energy representation and the level for each component $\hat{g}_j$.)

It is easy to use the construction (40) to produce the whole of the discrete series of Virasoro representations. To do this take $g = su(2) \oplus su(2)$ and $h$ to be the diagonal $su(2)$ subalgebra. For a level $x$ representation of $\widehat{su}(2)$,

$$c^{su(2)} = \frac{3x}{x+2}, \tag{47}$$

as $h^{su(2)} = 2$. If we take a representation of $\hat{g}$ formed by a level $m$ representation of the first $\widehat{su}(2)$ factor and a level 1 representation of the second, we obtain a level $m+1$ representation of $\hat{h}$. Then (42) becomes

$$\begin{aligned} c^K &= \frac{3m}{m+2} + 1 - \frac{3(m+1)}{(m+3)} \\ &= 1 - \frac{6}{(m+2)(m+3)}, \end{aligned} \tag{48}$$

giving the whole sequence (7).

## 8. Extended algebras

Just as the whole of the discrete series for the Virasoro algebra can be obtained from the coset construction for

$$su(2)_m \oplus su(2)_1 / su(2)_{m+1}, \tag{49}$$

where the suffices denote levels, so the whole of the discrete series for the super-Virasoro algebra can be obtained from the coset

$$su(2)_m \oplus su(2)_2 / su(2)_{m+2}. \tag{50}$$

It is easy to check that the formula (42) applied to (50) gives the discrete series (17) but to establish that we really have a construction

## 5. Extended Conformal Algebras

we must extend (40) to give expressions for $G_r$. To do this note that we can express the level two representation of $su(2)$ in terms of three real fermion fields, $\psi^i(z)$,

$$T^i(z) \equiv \sum_{n \in \mathbf{Z}} T_n^i z^{-n-1} = \frac{i}{2}\epsilon_{ijk}\psi^j(z)\psi^k(z). \tag{51}$$

If the generator of the level $m$ representation of $su(2)$ is denoted by $S^i(z)$, the fermionic generator of the superconformal transformations can be written [3]

$$G(z) = \frac{i}{6}\epsilon_{ijk}\psi^i(z)\psi^j(z)\psi^k(z) + \frac{2}{m}S^i(z)\psi^i(z). \tag{52}$$

This result can be generalised to construct extended algebras in other coset constructions [7], [8], providing in particular unitary representations of (20) corresponding to the coset [7]

$$so(3)_m \oplus su(3)_1 / so(3)_{m+4}. \tag{53}$$

Other examples of cosets providing extended conformal algebras are given in the table.

*Discrete Series of some Extended Conformal Algebras*

| algebra | $c$ | coset |
|---|---|---|
| Virasoro | $1 - \frac{6}{(m+2)(m+3)}$ | $\frac{su(2)_m \oplus su(2)_1}{su(2)_{m+1}}$ |
| Super-Virasoro | $\frac{3}{2}\left(1 - \frac{8}{(m+2)(m+4)}\right)$ | $\frac{su(2)_m \oplus su(2)_2}{su(2)_{m+2}}$ |
| $S_3$ parafermions | $2\left(1 - \frac{12}{(m+2)(m+6)}\right)$ | $\frac{so(3)_m \oplus su(3)_1}{so(3)_{m+4}}$ |
| $D_N$ parafermions | $(N-1)\left(1 - \frac{N(N-2)}{(m+N)(m_N-2)}\right)$ | $\frac{so(N)_m \oplus su(N)_1}{so(N)_{m+2}}$ |
| $Z_3$ parafermions | $2\left(1 - \frac{12}{(m+3)(m+4)}\right)$ | $\frac{su(3)_m \oplus su(3)_1}{su(3)_{m+1}}$ |
| $Z_N$ parafermions | $\frac{2(m-1)}{m+2}$ | $\frac{su(2)_m}{u(1)}$ |
| Casimir algebra | $r_g\left(1 - \frac{h^g(h^g+1)}{(m+h^g)(m+h^g+1)}\right)$ | $\frac{g_m \oplus g_1}{g_{m+1}}$ |
| affine $\hat{g}$ | $\frac{m \dim g}{m+h^g}$ | $g_m$ |
| superaffine $\hat{sg}$ | $\frac{m \dim g}{m+h^g} + \frac{1}{2}\dim g$ | $g_m \oplus g_{h^g}$ |

Many questions remain to be answered about the extended conformal algebras that we have been discussing. We would like to have results for them that parallel those for the Virasoro algebra and the super-Virasoro algebra. In particular, this means being able to classify their unitary highest weight representations, finding for which values of $c$ there are continua of representations and which values belong to discrete series. In the case of discrete series of unitary representations, we would want explicit coset constructions, from which the modular covariance of the characters would follow. More importantly one would wish to gain a deeper understanding of these modular properties, independent of the coset construction.

There are also many questions one can ask about the coset construction directly. For example, what is the extended conformal algebra associated with a given coset $g_x/h_y$ and under what circumstances does one have finite reducibility with respect to it? The cosets in the table are typically of the form $h_m \oplus g_x/h_{m+y}$. We can ask similar questions for such series and also try to understand when they really are discrete, and when continua of representations exist.

### References

[1] M.A. Virasoro, *Phys. Rev.* **D1** (1970)2933.

[2] D. Friedan, Z. Qiu and S. Shenker, *Phys. Rev. Lett.* **52** (1984)376; *Commun. Math. Phys.* **107** (1986)535.

[3] P. Goddard, A. Kent and D. Olive, *Phys. Lett.* **B152** (1985) 88; *Commun. Math. Phys.* **103** (1986)105.

[4] J.L. Cardy, *Nucl. Phys.* **B270** (1986)186.

[5] V.A. Fateev and A.B. Zamolodchikov, *Nucl. Phys.* **B280** (1987)644.

[6] H. Sugawara, *Phys. Rev.* **170** (1968)1659.

[7] P. Goddard and A. Schwimmer, *Phys. Lett.* **206B** (1988)62.

[8] P. Bowcock and P. Goddard, *Nucl. Phys.* **B3 05 [FS23]** (1988)685.

DEPARTMENT OF APPLIED MATHEMATICS AND THEORETICAL PHYSICS, SILVER STREET, CAMBRIDGE CB3 9EW, ENGLAND.

# 6

## Super Riemann Surfaces

*Alice Rogers*

A super Riemann surface is a particular kind of (1,1)-dimensional complex analytic supermanifold. From the point of view of supermanifold theory, super Riemann surfaces are interesting because they furnish the simplest examples of what have become known as non-split supermanifolds, that is, supermanifolds where the odd and even parts are genuinely intertwined, as opposed to split supermanifolds which are essentially the exterior bundles of a vector bundle over a conventional manifold. However undoubtedly the main motivation for the study of super Riemann surfaces has been their relevance to the Polyakov quantisation of the spinning string.

The starting point in this approach to the spinning string is a Euclidean version of Howe's formulation of the classical dynamics in (2,2)-dimensional superspace [1]. The action for the theory is

$$S = \frac{1}{4} \int d^2x d^2\theta E D_\alpha V D^\alpha V, \qquad (1)$$

where $V$ is a scalar function on a real (2,2)-dimensional supermanifold $N$. This supermanifold is constructed from a spin bundle of a real 2-dimensional manifold in the following standard way: suppose that $\{U_\alpha\}$ is a covering of M by trivialisation and coordinate neighbourhoods $U_\alpha$, that $\phi_\alpha : U_\alpha \to R^2$ are the coordinate functions and $g_{\alpha\beta} : U_\alpha \cap U_\beta \to U(1)$ are the transition functions of the spin bundle. Then the (2,2) dimensional supermanifold has coordinate transition functions

$$x \mapsto \phi_\alpha \circ \phi_\beta^{-1}(x) \qquad (2)$$

for the even coordinates (the new even coordinates thus do not depend on the old odd coordinates) and

$$\theta^i \mapsto g_{\alpha\beta}(x) {\gamma_5}^i{}_j \theta^j \qquad (3)$$

for the odd coordinates where ${\gamma_5}^i{}_j$ determines the spinor representation of $U(1)$. (Further details of this construction may be found in [2].) Incidentally, this is an example of a split supermanifold, indeed, by a theorem of Batchelor [3], all real smooth supermanifolds are split.

This construction of the (2,2)-dimensional supermanifold $N$ makes it clear that there is a reduction of the frame bundle of $N$ from the supergroup $GL(2,2)$ to the group $U(1)$. This reduced bundle is central to Howe's construction, and it is useful to consider 'vierbein' which are bases of the tangent space to $N$ consisting of two even vectors $E_a$ ($a = 1, 2$) and two odd vectors $E_\alpha$ ($\alpha = 1, 2$), transforming under $e^{i\theta} \in U(1)$ by the rule

$$\begin{pmatrix} E_a \\ E_\alpha \end{pmatrix} \mapsto \begin{pmatrix} V(e^{i\theta}) & 0 \\ 0 & S(e^{i\theta}) \end{pmatrix} \begin{pmatrix} E_a \\ E_\alpha \end{pmatrix}, \tag{4}$$

where $V$ and $S$ are respectively vector and spinor representations of $U(1)$. (Here and elsewhere lower case Roman indices will denote even, commuting objects, lower case Greek indices odd, anticommuting objects and upper case Roman indices will denote objects of either parity.) Already this tight structure on the frame bundle makes it clear that one is not simply considering Riemann geometry on a supermanifold. As in the superspace formulation of 4-dimensional supergravity [4], further structure is required to generate the correct physical theory. The next step is to choose a connection in the vierbein bundle such that the torsion $T_{AB}{}^C$, defined by

$$T_{AB}{}^C E^A E^B = T_A = DE_A, \tag{5}$$

where $E^A = (E^a, E^\alpha)$ are one-forms dual to $E_A = (E_a, E_\alpha)$, satisfies the constraints

$$\begin{aligned} T_{\alpha\beta}{}^c &= 2i\gamma^c{}_{\alpha\beta} \\ T_{\alpha\beta}{}^\gamma &= T_{ab}{}^c = 0. \end{aligned} \tag{6}$$

These constraints mean that the various components $E_A{}^M$ of the vierbein cannot be freely specified; it is shown by Howe [1] that these constraints do lead to the desired physical theory of the spinning string. (There does not seem to be a fundamental method for deriving the constraints, they are justified by the fact that they lead to the correct physics.) Although the (2,2)-dimensional geometry is somewhat messy, it is essential as a bridge between the 'real' world of physics and the more elegant world of super Riemann surfaces.

Polyakov quantisation involves integration over the space of all $E_M{}^A$ (subject to the constraints) and all superfields $V$, modulo the symmetries of the theory. As will now be explained, this space can in fact be identified with the supermoduli space of super Riemann structures. The symmetries of the theory are the group of general coordinate transformations of the supermanifold $N$, and a group of transformations of the vierbein defined by Howe which are superspace

## 6. Super Riemann Surfaces

analogues of the conformal transformations of a conventional Riemannian metric. These transformations, known as super-Weyl transformations, act on the vierbein in the following manner:

$$E_M{}^a \mapsto \Lambda E_M{}^a$$
$$E_M{}^\alpha \mapsto \Lambda^{\frac{1}{2}} E_M{}^\alpha - \tfrac{i}{2}\Lambda^{-\frac{1}{2}} \gamma_a{}^{\alpha\beta} D_\beta \Lambda. \tag{7}$$

(Here the parameter $\Lambda$ is an even invertible scalar function on the supermanifold $N$. These transformations, which were first derived by Howe, are slightly more complicated than a simple rescaling of the vierbein, but are the simplest which preserve the torsion constraints while placing no restrictions on the parameter $\Lambda$.) Thus one finds that the Polyakov quantisation of the spinning string requires one to integrate over the space of super-Weyl and supersmooth equivalence classes on the real (2,2)-dimensional smooth supermanifold $N$, $N$ itself being completely determined by the genus of the underlying manifold $M$ together with the choice of spin structure on $M$. An important result of Howe [1], that a local coordinate system can always be chosen in which the vierbein are super-Weyl flat, provides the link between the (2,2)-dimensional real geometry and the (1,1)-dimensional super Riemann surface concept developed by Baranov and Schwarz [5] and by Friedan [6]. (A vierbein $E_M{}^A$ is said to be super-Weyl flat if it is obtainable from the flat super vierbein

$$\bar{E}_m{}^a = \delta_m^a \qquad \bar{E}_m{}^\alpha = 0$$
$$\bar{E}_\mu{}^a = i\theta^\lambda \gamma^a{}_{\lambda\mu} \qquad \bar{E}_\mu{}^\alpha = \delta_\mu{}^\alpha \tag{8}$$

by a super Weyl transformation. That is, $E_M{}^A$ is super Weyl flat if it is of the form

$$E_M{}^a = \Lambda \bar{E}_M{}^a$$
$$E_M{}^\alpha = \Lambda^{\frac{1}{2}} \bar{E}_M{}^\alpha - i\bar{E}_M{}^a \gamma_a{}^{\alpha\beta} \bar{D}_\beta \Lambda^{\frac{1}{2}} \tag{9}$$

for some parameter $\Lambda$.)

Now a super Riemann surface [5,6] is defined to be a (1,1) dimensional complex analytic supermanifold with changes of local coordinates $(z, \zeta) \mapsto (\tilde{z}, \tilde{\zeta})$ restricted by the requirement that they are superconformal, that is that the differential operator

$$D := \frac{\partial}{\partial \zeta} + \zeta \frac{\partial}{\partial z} \tag{10}$$

transforms covariantly with

$$D = (D\tilde{\zeta})\tilde{D}. \tag{11}$$

The relationship between superconformal classes of super Riemann surfaces and super-Weyl equivalence classes of (2,2)-dimensional real supermanifolds based on surfaces with spin-structures then emerges in the following way. Starting with the real (2,2)-dimensional supermanifold $N$, and using Howe's result, one chooses everywhere coordinate systems such that the vierbein $E_M{}^A$ are super-Weyl flat. Putting the super-Weyl flat coordinate systems $(x, y, \theta_1, \theta_2)$ into complex form $(z, \bar{z}, \zeta, \bar{\zeta})$ with $z = x + iy$ and $\zeta, \bar{\zeta}$ chiral, one finds that allowed changes of coordinate, which preserve the super-Weyl flatness of the vierbein, are superconformal, and conversely super-Weyl transformations preserve the super-Weyl flatness [7] and [8]. Thus a super Riemann structure has been obtained. As a result, one finds that super-Weyl structures on $N$ are in one-to-one correspondence with super Riemann structures, and thus one may investigate the space of all super-Weyl structures on $N$ by investigating the corresponding supermoduli space of super Riemann surfaces. A detailed study of supermoduli space was initiated by Crane and Rabin [9], and is still continuing.

The simplest kind of super Riemann surface is constructed from a Riemann surface $S$ together with a spin structure on $S$. Noting that the defining conditions (11) of a super Riemann surface mean that a change of coordinates has the form

$$\begin{aligned}\tilde{z}(z, \zeta) &= f(z) + \zeta\psi\sqrt{f'(z)} \\ \tilde{\zeta}(z, \zeta) &= \psi(z) + \zeta\sqrt{f'(z) + \psi(z)\psi'(z)},\end{aligned} \quad (12)$$

where $f$ and $\psi$ are respectively even and odd analytic functions, one can use a standard patching construction to obtain a split super Riemann surface by setting $\psi$ to be zero, and letting $f$ simply be the transition function of the Riemann surface $S$, with the sign of the square roots determined by the spin structure. Following Crane and Rabin [9], such a super Riemann surface will be called the canonical Riemann surface over $S$ corresponding to the given spin structure. Not all super Riemann surfaces are canonical; if this were the case, supermoduli space would in fact simply be spin moduli space, whereas in fact supermoduli space is a superspace with anticommuting as well as commuting coordinates.

The first step in the investigation of supermoduli space is the uniformisation theorem established by Crane and Rabin [9]. They first prove that there are only three simply connected super Riemann surfaces. (For a given $(m, n)$-dimensional supermanifold $N$, there is a natural projection onto an underlying $m$-dimensional conventional manifold $N_0$. A supermanifold is said to be simply connected if that

## 6. Super Riemann Surfaces

is the case for its underlying manifold.) These simply connected manifolds are the canonical super Riemann surfaces over the three simply connected Riemann surfaces $C$ (the complex plane), $C_*$ (the extended complex plane) and $U$ (the upper half plane), each with its unique spin structure. These super Riemann surfaces are denoted $SC$, $SC_*$ and $SU$ respectively. The uniformisation result is proved by considering a general super Riemann surface over $C$, $C_*$ or $U$, and showing, using cohomological arguments, that one can unpick the coordinate transition functions level by level in the Grassmann algebra until they become of canonical form. Consideration of covering spaces then shows that any super Riemann surface is the quotient of either $SC$, $SC_*$ or $SU$ by a discrete subgroup $\Gamma$ of superconformal automorphisms acting properly discontinuously with respect to the De Witt topology. (This topology is defined in [10] and described in [9].) The fundamental group of a super Riemann surface $T$ is the same as the fundamental group of the underlying Riemann surface $T_0$, and isomorphic to $\Gamma$. Considering the case of genus $g > 1$, transformations of $SU$ of the form (12), with

$$f(z) = \frac{az+b}{cz+d} \text{ and } \psi(z) = \frac{\gamma z + \delta}{cz+d} \tag{13}$$

(where $a, b, c, d, \gamma$ and $\delta$ are real and $ab - cd = 1$), are superconformal automorphisms. Other superconformal automorphisms also exist, but do not lead to distinct superconformal structures as was shown by Hodgkin [11]. Thus, using the fact that the fundamental group of a surface of genus $g$ has $2g$ generators and one relation, one finds that supermoduli space has real dimension $(6g - 6, 4g - 4)$. Of course this approach simply gives a rough dimension count, and does not determine the global nature of supermoduli space. Further work on supermoduli space and super Teichmuller space has been carried out by various authors [12], and recent work by Bryant [13] suggests that supermoduli space may be split.

Recalling that the Polyakov quantisation of the spinning string involves integration over supermoduli space, the splitness or otherwise of supermoduli space assumes considerable importance, because it is often not possible to construct a well-defined integral on a non-split supermanifold using standard techniques if the supermanifold has a nontrivial boundary. The following simple example of integration on a (1,2)-dimensional supermanifold [14] demonstrates the difficulty that may occur. According to the standard rules for superspace integration [15], if $x$ is the even co-ordinate and $\theta^1$ and $\theta^2$ are the odd co-ordinates,

$$\int_0^1 dx\, d\theta^1\, d\theta^2\, x = 0. \tag{14}$$

However, if one now makes changes to new co-ordinates $y, \phi^1, \phi^2$ with

$$y = x + \theta^1\theta^2, \quad \phi^1 = \theta^1, \quad \phi^2 = \theta^2, \tag{15}$$

one finds (using the Berezin transformation rule) that the measures are unchanged, and thus evaluation of the integral in the transformed co-ordinates gives

$$\int_0^1 dx\, d\phi^1\, d\phi^2 (y - \phi^1\phi^2) = -1, \tag{16}$$

which is clearly inconsistent. This problem does not occur if the integrand is sufficiently smooth, and either the integration region is compact without boundary or the integrand is zero on the boundary, but this is not generally the case for the integrals involved in superstring quantisation. Inconsistencies are also avoided if all changes of co-ordinate $(x, \theta) \mapsto (y, \phi)$ are such that all partial derivatives $\frac{\partial y^i}{\partial \theta^j}$ are zero, as is the case for split supermanifolds—and thus it might be possible to define a consistent superstring quantisation if supermoduli space were to be split; however for this to be the case the splitting would have to be canonical in some sense.

If supermoduli space does not admit a suitably canonical splitting, there is another possible method, due to Batchelor, Penkov and Rothstein [16] [17] [18], which leads to a consistent integral. In this approach to integration on an $(m, n)$-dimensional supermanifold, instead of regarding $dx^1 \ldots dx^m\, d\theta^1 \ldots d\theta^n$ as a composite volume form transforming by a factor equal to the super Jacobian, one regards it as the form-valued differential operator $dx^1 \ldots dx^m \frac{\partial}{\partial \theta^1} \ldots \frac{\partial}{\partial \theta^n}$. This leads naturally to a consistent transformation rule for integrals, but to apply it to superstrings [19] one would have to construct a suitable form-valued differential operator out of the material available in the spinning string formalism.

To end this lecture, an alternative approach to integration on $(1,1)$-dimensional supermanifolds, which does not lead to boundary ambiguities, will be described. This leads to a well-defined contour integral on a super Riemann surface, and it is also possible that higher-dimensional analogues might provide a consistent integral on supermoduli space. Suppose that $t, \tau$ are co-ordinates on the trivial $(1,1)$-dimensional real supermanifold $R^{1,1}$, that $g$ is a function on $R^{1,1}$ and

## 6. Super Riemann Surfaces

that $(a,\alpha)$ and $(b,\beta)$ are points in $R^{1,1}$. Then the integral between limits $(a,\alpha)$ and $(b,\beta)$ is defined to be

$$\int_\alpha^\beta \int_a^b g(t,\tau)\,DT := \int \left[\int_{a+\tau\alpha}^{b+\tau\beta} g(t,\tau)\,dt\right] d\tau. \tag{17}$$

An important property of this integral is that it obeys what might be called the square root of the fundamental theorem of calculus, that is, if $D_T$ denotes the superdifferentiation operator $\frac{\partial}{\partial \tau} + \tau \frac{\partial}{\partial t}$ (which satisfies $D^2 = \frac{\partial}{\partial t}$), then

$$\int_\alpha^\beta \int_a^b D_T h(t,\tau)\,DT = h(b,\beta) - h(a,\alpha). \tag{18}$$

Additionally, this integral has the following consistent transformation rule: suppose that

$$\begin{aligned} K: R^{1,1} &\to R^{1,1} \\ (t,\tau) &\mapsto (s(t,\tau),\sigma(t,\tau)), \end{aligned} \tag{19}$$

with $K$ 'real superconformal', that is, of the form

$$\begin{aligned} s(t,\tau) &= a(t) + \tau\alpha(t)\sqrt{a'(t)} \\ \sigma(t,\tau) &= \alpha(t) + \sqrt{a'(t) + \alpha(t)\alpha'(t)}. \end{aligned} \tag{20}$$

Then, if $(c,\gamma) = K(a,\alpha)$ and $(d,\delta) = K(b,\beta)$,

$$\int_\gamma^\delta \int_c^d g(s,\sigma)\,DS = \int_\alpha^\beta \int_a^b g(K(t,\tau))D_T K(t,\tau)\,DT. \tag{21}$$

This result follows from equation (18) together with the chain rule for superdifferentiation of combinations of superconformal functions.

This (1,1) dimensional integral leads naturally to a theory of contour integration on super Riemann surfaces [20]. Starting with $SC$ or $C^{1,1}$, the trivial super Riemann surface, one defines a contour to be a mapping

$$\begin{aligned} C: R^{1,1} &\to C^{1,1} \\ (t,\tau) &\mapsto (c(t,\tau),\gamma(t,\tau)) \end{aligned} \tag{22}$$

that is real superconformal. Then, if $C(a,\alpha) = (p,\pi)$ and $C(b,\beta) = (r,\rho)$, the integral of a function $f$ on $C^{1,1}$ between $(p,\pi)$ and $(r,\rho)$ is defined to be

$$\int_{C[p,\pi;r,\rho]} f(z,\zeta)\,DZ := \int_\alpha^\beta \int_a^b f(c(t,\tau),\gamma(t,\tau))D_T(\gamma(t,\tau))\,DT. \tag{23}$$

It follows from the transformation rule (21) for real (1,1)-dimensional integrals that this integral is reparametrisation invariant. Additionally, further use of the chain rule shows that it has a natural transformation rule under superconformal changes of co-ordinate on $C^{1,1}$. Specifically, if

$$F : C^{1,1} \to C^{1,1}$$
$$(z,\zeta) \mapsto (\tilde{z}(z,\zeta),\tilde{\zeta}(z,\zeta)) \qquad (24)$$

is superconformal, and $F(p,\pi) = (p',\pi')$, $F(r,\rho) = (r',\rho')$, then

$$\int_{F \circ C[p',\pi';r',\rho']} g(\tilde{z},\tilde{\zeta}) D\tilde{Z} = \int_{C[p,\pi;r,\rho]} g(F(z,\zeta)) D_Z \tilde{\zeta} \, DZ. \qquad (25)$$

Now on a super Riemann surface one can define a half-form as an object which is written in local co-ordinates as $W(Z) DZ$ with transformation rule $W(Z) = W(\tilde{Z}) D_Z \tilde{\zeta}$. Thus the contour integral defined above leads to a consistent method for integration of half-forms over contours on super Riemann surfaces.

A natural generalisation of the standard result leads to a Cauchy theorem for these contour integrals, and also a contour integral representation for analytic functions, which takes the form

$$f(r,\rho) = \frac{2\pi i}{n(C;r,\rho)} \int_C f(z,\zeta) \frac{(\zeta - \rho)}{(z - r)} \, DZ,$$
$$\text{where } n = \frac{1}{2\pi i} \int_C \frac{(\zeta - \rho)}{(z - r)} \, DZ. \qquad (26)$$

Further details may be found in [17].

A large number of papers have been written on super Riemann surfaces, and it is not possible to do them all justice in this brief survey. Recent work by Rabin and Topiwala [21] has demonstrated that all super Riemann surfaces are algebraic, and this makes possible the use of a whole battery of techniques from algebraic geometry. Some areas of considerable difficulty remain; as well as the problems associated with integration over moduli space, the interesting discussion of line bundles given by Giddings and Nelson [22] shows that the space of sections of certain line bundles is liable to jump in dimension along trajectories in moduli space, and is not a free module over the Grassmann algebra. (Further discussion of this point has been given by Hodgkin [23].) As well as being mathematically disappointing, these properties can lead to difficulties in the Polyakov quantisation of the spinning string, where dimension counting of such spaces is used. The author is currently considering the rôle of an alternative kind of bundle on super Riemann surfaces as a means of resolving this difficulty.

## Acknowledgements

The author is grateful to the Science and Engineering Research Council of Great Britain for financial support under Advanced Research Fellowship B/AF/687.

## References

[1] P.S. Howe, Super Weyl Transformations in Two Dimensions, *J. Phys.* **A12** (1979)393–402.

[2] A. Rogers, Super Lie groups: global topology and local structure, *J. Math. Phys.* **22** (1981)939–945.

[3] M. Batchelor, The Structure of Supermanifolds, *Trans. Am. Math. Soc.* **253** (1979)329–338.

[4] J. Wess and B. Zumino, Superspace Formulation of Superspace Supergravity, *Phys. Lett.* **B66** (1977) 361.

[5] M. Baranov and A. Schwarz, Multiloop Contributions in String Theory, *Pisma ZETF* **42**(1985) 340[=*JETP letters* **42** (1986)419].

[6] D. Friedan, Notes on String Theory and Two Dimensional Conformal Field Theory, in *Proceedings of the Workshop on Unified String Theories*, ed. D. Gross and M. Green, World Scientific Press, Singapore, 1986.

[7] B. Zumino, Supersymmetry, *Proc.Conf. at Northeastern University, 1975*, eds. R. Arnowitt and P. Nath, MIT Press, Cambridge (USA) and London (UK).

[8] S.B. Giddings and P. Nelson, The Geometry of Super Riemann Surfaces, *Comm. Math. Phys.* **116** (1988)607–634.

[9] L. Crane and J. Rabin, Super Riemann Surfaces: Uniformization and Teichmuller Theory, *Comm. Math. Phys.* **113** (1988)601–623.

[10] B.S. De Witt, *Supermanifolds*, Cambridge University Press, Cambridge, 1984.

[11] L. Hodgkin, A Direct Calculation of Super Teichmuller Space, *Lett. Math. Phys.* **14** (1987)57–63.

[12] For example: C. Le Brun and M. Rothstein, Moduli of Super Riemann Surfaces, *Comm. Math. Phys.* **117** (1988)159–176; S. Uehara and Y. Yasui, Super Beltrami differentials via 2$D$ supergravity, *CERN preprint* TH.5031/88.

[13] P.A. Bryant, Structure of supermoduli space, I,II and III, *Preprints*, Cambridge, August 1988.

[14] D.A. Lêites, Introduction to the Theory of supermanifolds, *Uspheki Mat. Nauk.* **35** (1980)3–57 [=*Russ. Math. Surv.* **35** (1980)1–64].

[15] F.A. Berezin, *The method of second quantisation*, Academic Press, New York, 1966.

[16] M. Batchelor, Algebraic integration, *unpublished*.

[17] I.B. Penkov, $\mathcal{D}$-modules on Supermanifolds, *Invent. Math.* **71** (1983)501–512.

[18] M. Rothstein, Integration on Noncompact Supermanifolds, *Trans. Am. Math. Soc.* **289** (1987)387–396.

[19] M. Martellini and P. Teofillato, Global structure of the superstring partition function and resolution of the supermoduli measure ambiguity, *King's College preprint*, February 1988.

[20] A. Rogers, Contour integration on super Riemann surfaces, *Phys. Lett.* **B213** (1988)37–40.

[21] J. Rabin and P. Topiwala, Super Riemann surfaces are algebraic curves, *UCSD preprint*, June 1988.

[22] S.B. Giddings and P. Nelson, Line Bundles on Super Riemann surfaces, *Comm. Math. Phys.* **18** (1988)289–302.

[23] L. Hodgkin, Problems of Fields on Super Riemann Surfaces, *King's College London Preprint* 1988, to appear in *Journ. Geom. in Physics*.

DEPARTMENT OF MATHEMATICS, KING'S COLLEGE, STRAND, LONDON WC2R 2LS, ENGLAND.

# 7

## Gauge Theories and Relativistic Membranes

*P.K. Townsend*

In the course of its time evolution, a relativistic point particle traces out a "worldline" $w$ in Minkowski spacetime $(M,\eta)$. This provides us with an immersion of $w$ into $M$, by virtue of which the Minkowski metric $\eta$ of $M$ induces a metric on $w$. For a segment of the worldline with endpoints $A$ and $B$ we take the particle's action, $S$, to be the length, in the induced metric, of the segment. In local coordinates $\{X^\mu,\ \mu = 0, 1, \ldots, (d-1)\}$ for $M$, and arbitrary local coordinate $t$ for $w$, the action is

$$S = -m \int_A^B dt\, [-\dot{X}^\mu \dot{X}^\nu \eta_{\mu\nu}]^{\frac{1}{2}}. \tag{1}$$

The constant $m$ is (in units where the speed of light is unity) the particle's mass. The Euler-Lagrange equations that follow from this action imply that $w$ is a geodesic.

In 1962 Dirac introduced the generalization of this action for a relativistic membrane, i.e. a two-dimensional extended object [1]. In the course of its time evolution it sweeps out a three-dimensional "worldvolume" $W$. If we suppose the locus of the membrane to be a closed two-dimensional manifold then $W$ will be a three-manifold with boundaries, which we shall again denote by $A$ and $B$, at the initial and final times. If $\xi^i$ are local coordinates on $W$ then

$$S_D = -T \int_A^B d^3\xi\, [-\det(\partial_i X^\mu \partial_j X^\nu \eta_{\mu\nu})]^{\frac{1}{2}}, \tag{2}$$

where $T$ is the surface tension of the membrane, (the "mostly plus" metric convention will be used throughout this article). The further generalization to a $p$-dimensional extended object, which sweeps out a $(p+1)$-dimensional worldvolume is obvious. A particularly interesting case is $p = 1$ for which the action is that of Nambu and Goto for a relativistic string. Of interest here will be the relativistic membrane with action (2) because of its remarkable connection with Yang-Mills gauge theories [2]. This connection has been extended recently [3] to supermembranes [4]. Some simplifications, and generalisations, of these results have been obtained in a recent work [5], on which the presentation here is based. I shall begin, however, by examining some other intriguing connections between membranes and gauge theories.

First is a similarity between membranes and general relativity. In order to make this clearer let us adopt the same notation for the curved spacetime of general relativity as for the worldvolume of an extended object. Thus $\{\xi^i,\ i = 0, 1, \ldots, (n - 1)\}$ will be local coordinates for an $n$-dimensional manifold furnished with a symmetric connection $\Gamma^i{}_{jk}$. From the curvature tensor $R^k{}_{ilj}$ for $\Gamma$ we can construct the symmetric Ricci tensor $R_{ij}(\Gamma) = \frac{1}{2}(R^k{}_{ikj} + R^k{}_{jki})$. Then, even *without* a metric, the action

$$S_{ES} = -T \int d^n\xi \, [-\det R_{ij}(\Gamma)]^{\frac{1}{2}} \tag{3}$$

analogous to (2) is available. This is the Eddington-Schrödinger action, and it is equivalent to the Einstein-Hilbert action with the addition of a cosmological term. I shall explain the equivalence here because it involves a trick that will be needed again for the membrane. Although the action (3) does not involve a metric, we can introduce one as an auxiliary, non-dynamical, variable. Consider the action

$$S = -\frac{1}{2\kappa} \int d^n\xi \, [\sqrt{-\gamma}\gamma^{ij} R_{ij}(\Gamma) - (n-2)\Lambda\sqrt{-\gamma}], \tag{4}$$

with $\kappa$ a constant with dimensions of $(\text{length})^{(n-2)}$, $\Lambda$ the cosmological constant, and $\gamma = \det \gamma_{ij}$. The equation of motion for $\gamma_{ij}$ is

$$R_{ij}(\Gamma) = \Lambda \gamma_{ij}, \tag{5}$$

which can be considered as an equation determining the metric in terms of the connection. Substituting $\Lambda^{-1} R_{ij}$ for $\gamma_{ij}$ in (4) we recover the action (3) on identifying $T$ as $(\kappa^2 \Lambda^{(n-2)})^{-\frac{1}{2}}$. On the other hand, the field equation for $\Gamma$ is

$$\Gamma^i{}_{jk} = \begin{Bmatrix} i \\ jk \end{Bmatrix}, \tag{6}$$

i.e. the Levi-Civita connection, and substituting this in (4) produces the usual Einstein-Hilbert action with the cosmological term proportional to $\Lambda\sqrt{-\gamma}$. This establishes the (classical) equivalence of (3) with (4). Let us note that equation (5) can be written as

$$\mathcal{R}_{ij} = 0, \tag{7}$$

where $\mathcal{R}_{ij}$ is the symmetrized contraction of

$$\mathcal{R}^k{}_{ilj} \equiv R^k{}_{ilj} - \frac{2\Lambda}{(n-1)}(\delta^i{}_{(i}\eta_{j)k}),$$

which can be viewed as the "field strength" tensor of the three-dimensional de Sitter group $SO(3,1)$ if $\Lambda > 0$, or of the anti-de Sitter group $SO(2,2)$ if $\Lambda < 0$ [6].

Using the same auxiliary metric trick (introduced in the membrane context by Howe and Tucker [7]), we can rewrite the action (2) as

$$S_{HT} = -\frac{T}{2} \int d^3\xi \, [\sqrt{-\gamma}\gamma^{ij}\partial_i X^\mu \partial_j X^\nu \eta_{\mu\nu} - \sqrt{-\gamma}], \qquad (8)$$

which has the appearance of a "non-linear $\sigma$-model", with $(M,\eta)$ as target space, in a background gravitational field $\gamma_{ij}$, and $\sqrt{-\gamma}$ a "cosmological" term. The equation of motion for $X^\mu$ is

$$\gamma^{ij}\nabla_i \partial_j X^\mu = 0, \qquad (9)$$

where $\nabla$ is a covariant derivative.

Now just as the equation (7) tells us that the "trace" (i.e. contraction) of the *intrinsic* curvature vanishes so the equation (9) tells us that the trace of the *extrinsic* curvature, given by the second fundamental forms

$$\pi^\mu_{ij} = \nabla_i \partial_j X^\mu, \qquad (10)$$

vanishes [8].

These similarities between general relativity and membranes are even more striking when we compare *three*-dimensional general relativity with a membrane in a *three* dimensional spacetime. The *kinematics* of the two are then identical. The dynamics differ but in an interesting way. Let us introduce gauge field one-forms

$$e^\mu = d\xi^i e_i{}^\mu \qquad \omega^\mu = d\xi^i \omega_i{}^\mu \qquad (11)$$

for the (Hermitian) generators $P_\mu$ and $M_\mu$ of the three-dimensional anti de Sitter group $SO(2,2) \cong SO(2,1) \times SO(2,1)$. Then introducing a metric $\gamma$ by

$$\gamma_{ij} = e_i{}^\mu e_j{}^\nu \eta_{\mu\nu} \qquad (12)$$

the action (4) is easily shown to be equivalent to the integral of the three-form

$$\Omega_- = e^\mu d\omega^\nu \, \eta_{\mu\nu} + a e^\mu \omega^\nu \omega^\rho \, \varepsilon_{\mu\nu\rho} + b e^\mu e^\nu e^\rho \, \varepsilon_{\mu\nu\rho}, \qquad (13)$$

where $a$ and $b$ are two constants which can be changed by rescaling $e^\mu$, $\omega^\mu$, and $\Omega_-$. The three-form $\Omega_-$ is, in fact, the difference of the two Chern-Simons 3-forms for the two $SO(2,1)$ factors of $SO(2,2)$ [9], [10], [11]. Now if we rescale $b$ to zero we simply get three dimensional general relativity without a cosmological term. If, on the other hand, we rescale $a$ to zero we get the action for a membrane! To see this we observe that when $a = 0$ the equation of motion for $\omega$ is

$$de^\mu = 0, \qquad (14)$$

which we can obviously solve, at least locally, by setting

$$e^\mu = dX^\mu. \qquad (15)$$

With appropriate normalisation the remaining term in the action is then

$$-T\int_W \varepsilon_{\mu\nu\rho} dX^\mu dX^\nu dX^\rho = -T\int_W d^3\xi \, \det(\partial_i X^\mu), \qquad (16)$$

which clearly depends only on the *boundaries* $A$ and $B$ of $W$.

Now for a three-dimensional spacetime $\partial_i X^\mu$ are the entries of a $3 \times 3$ matrix $(\partial X)$ and so the action (2) is

$$S_D = -T\int_W d^3\xi [-\det\{(\partial X)(\eta)(\widetilde{\partial X})\}]^{\frac{1}{2}} \qquad (17)$$
$$= -T\int_W d^3\xi |\det(\partial_i X^\mu)|,$$

which differs from (16) only by the fact that we have the modulus of the determinant in (17) rather than the determinant. This difference is without significance at the classical level and so we may consider the action (2) to be equivalent to that of (16). (Quantum mechanically they are presumably not equivalent, so we would have to choose between them when passing to the quantum theory). We conclude, therefore, that the three-dimensional membrane is a type of "topological" field theory, an observation that was made in [12] in the context of other analogies between membranes and gauge theories.

This result is not be too surprising because a three-dimensional membrane, like three-dimensional general relativity, is "trivial" in the sense that a naive count of the degrees of freedom would lead to the conclusion that there are none. In fact, as emphasised by Witten [11] for the case of three-dimensional gravity, there may remain a finite number of degrees of freedom. An analogous result has been claimed by Fujikawa [13] for $p$-dimensional extended objects in a $(p+1)$ dimensional spacetime, which includes (for $p=2$) the case considered here.

## 7. Gauge Theories and Relativistic Membranes

In this "Chern-Simons" formulation of the three-dimensional membrane the action is

$$S = -3T \int_W [\eta_{\mu\nu}\omega^\mu de^\nu + \frac{1}{3}\varepsilon_{\mu\nu\rho}e^\mu e^\nu e^\rho], \tag{18}$$

and the field equations are equivalent to the vanishing of the field strength tensor for the gauge field one-forms associated with the Lie algebra

$$[M_\mu, M_\nu] = 0 \qquad [M_\mu, P_\nu] = 0$$
$$[P_\mu, P_\nu] = i\eta^{\rho\sigma}\varepsilon_{\mu\nu\rho}M_\sigma \tag{19}$$

which is a contraction of the Lie algebra of $SO(2,2)$. In this sense the action (18) is a Chern-Simons term for the algebra (19) (although the Killing metric of the algebra (19) is zero). A similar group contraction may be possible for the infinite dimensional extensions of $SO(2,2)$ considered by Blencowe [10]. Recently it has been shown [14] that these extensions are closely related to the group of symplectic diffeomorphisms of a two-dimensional surface, a group that will play a central role in the remainder of this article.

We return now to the action of (8) for a membrane in a $d$-dimensional spacetime. Because of the three-dimensional general coordinate invariance of this action not all of the variables are physically relevant. If we wish to express the action in terms of physically relevant variables only we shall have to fix a gauge. We suppose that the worldvolume is foliated by spacelike surfaces, the consecutive configurations of the membrane at worldvolume times $\xi^0 \equiv t$. Let $h_{ab}(\sigma, t)$ be the metric, in local coordinates $\{\xi^a \equiv \sigma^a, a = 1, 2\}$, induced by $\gamma_{ij}$ on these surfaces. Then the worldvolume 3-metric can be parametrized as

$$\gamma_{ij} = \begin{pmatrix} \phi^{-2}(-h + h_{cd}u^c u^d) & -\phi^{-1}h_{bc}u^c \\ -\phi^{-1}h_{ac}u^c & h_{ab} \end{pmatrix}, \tag{20}$$

where $h = \det h_{ab}$, and the variables, $\phi$, $u^a$, are related to the lapse and shift functions of the Hamiltonian formulation of general relativity. With this parametrization we have that

$$\sqrt{-\gamma} = \phi^{-1}h. \tag{21}$$

We also define

$$X^\pm = \frac{1}{\sqrt{2}}(X^0 \pm X^{d-1}), \tag{22}$$

and denote the remaining "transverse" variables by

$$\{X^I, I = 1, \ldots, (d-2)\}.$$

We now make the partial gauge choice

$$X^+ = t \qquad \phi = 1, \tag{23}$$

which we shall call the "light-cone gauge". Strictly speaking we may only set $\dot{X}^+ = p^+(t)$ where $p^+$ is a residual variable that is subsequently required to be constant by the field equations. This constant cannot necessarily be set to unity but, for the sake of simplicity, I shall do so here and refer to [5] for a fuller treatment.

Having made this gauge choice the field equation for $h_{ab}$ becomes

$$h_{ab} = \partial_a X^I \partial_b X^I, \tag{24}$$

(summation over $I$ being understood), and we shall use this in what follows.

At this stage the remaining variables are $(X^I, X^-, u^a)$. Substitution of (23) and (24) into the action (8) yields

$$S = \frac{T}{2} \int dt \int d^2\sigma \, [(\mathcal{D}_0 X^I)^2 - \det(\partial_a X^I \partial_b X^I) - 2X^- \partial_a u^a], \tag{25}$$

where

$$\mathcal{D}_0 = \partial_0 + u^a \partial_a \tag{26}$$

plays the role of a covariant time derivative.

It is sometimes stated that one cannot legitimately substitute gauge conditions into the action, and that information is lost in so doing. This is not so, however. Provided the gauge choice is *legitimate*, i.e. can be achieved by a gauge transformation, and does not, for example, involve the use of field equations in any way, then the "lost" information is redundant. The usual treatment of the light-cone gauge in string theory is not legitimate in this sense, whereas the gauge choice (23) (which applies equally well to a string as to a membrane) is.

Observe that the $X^-$ variable appears in the action (25) as a Lagrange multiplier for the constraint $\partial_a u^a = 0$. For a membrane of spherical topology, for which the first homology group is trivial, this constraint can be solved by

$$u^a = \varepsilon^{ab} \partial_b \omega. \tag{27}$$

This allows us to write $\mathcal{D}_0 X^I$ as

$$\mathcal{D}_0 X^I = \dot{X}^I - [\omega, X^I]_{lb}, \tag{28}$$

where we have introduced the *Lie bracket*

$$[f, g]_{lb} = \varepsilon^{ab} \partial_a f \, \partial_b g \tag{29}$$

## 7. Gauge Theories and Relativistic Membranes

for any two differentiable functions on the membrane. The potential term in the action can also be expressed in terms of this Lie bracket, so the final "light-cone action" becomes

$$S = \int dt \int d^2\sigma \left\{ \tfrac{1}{2}(\dot{X}^I - [\omega, X^I]_{lb})(\dot{X}^I - [\omega, X^I]_{lb}) \right. \tag{30}$$
$$\left. - \tfrac{1}{4}[X^I, X^J]_{lb}[X^I, X^J]_{lb} \right\}.$$

To see the significance of this result consider, for purposes of comparison, a Yang-Mills gauge theory in a $(d-1)$-dimensional flat spacetime for which space is a $(d-2)$-torus $T^{(d-2)}$ of unit volume. The action is

$$S = -\frac{1}{4}\int dt dx^{(d-2)} \, \text{tr}(F_{\mu\nu}F^{\mu\nu}), \tag{31}$$

where $F_{\mu\nu} = \partial_\mu A_\nu - \partial_\nu A_\mu + [A_\mu, A_\nu]$ is the Yang-Mills field strength for the gauge potential $A_\mu$ in the adjoint representation of a compact Lie algebra, and "tr" denotes the trace of matrices in this representation. We can expand the components $A^\mu = (A^0, A^I)$ of the gauge potential in harmonics on $T^{(d-2)}$. By keeping only the zero modes in this expansion, which is equivalent to setting $\partial_I = 0$, we effect a "dimensional reduction" to one dimension (time). In this way the action (31) reduces to

$$S = \int dt \, \text{tr}\left\{\frac{1}{2}(\partial_0 A^I - [A^0, A^I])(\partial_0 A^I - [A^0, A^I]) - \frac{1}{4}[A^I, A^J][A^I, A^J]\right\}. \tag{32}$$

By the replacement

$$A^0 \mapsto \omega \qquad A^I \mapsto X^I$$
$$[\,,\,] \mapsto [\,,\,]_{lb} \qquad \text{tr} \mapsto \int d^2\sigma \tag{33}$$

we recover the action (30), for which the gauge group can clearly be identified as the subgroup of the diffeomorphism group of the membrane that preserves the Lie bracket. This is the *area-preserving* or *symplectic diffeomorphism group*, SDiff. Thus the relativistic membrane is equivalent to a one-dimensional gauge theory of the infinite dimensional group SDiff. For the case of a spherical membrane, as assumed above, Hoppe has shown [2] that this group is isomorphic to $SU(\infty)$( strictly $SU_+(\infty)$).

Much of the mathematics of membranes resembles that of two-dimensional fluid dynamics. The derivative $\mathcal{D}_0$ is the convective derivative for a fluid with velocity field $u^a$ and $\partial_a u^a = 0$ is the condition for incompressible flow. The corresponding stream function $\omega$ here plays the role of a gauge field. The group SDiff also plays an important role in the fluid dynamics of incompressible flow in two dimensions; Arnold has shown that Euler's equations for the motion of a perfect fluid are equivalent to geodesic motion, in a left-invariant metric, on the group SDiff [15]. In a sense, therefore, the relativistic membrane and the dynamics of a perfect fluid are complementary systems.

The Hamiltonian associated with the action (30) is

$$H = \int d^2\sigma \left(\frac{1}{2}P^I P^I + \omega[P^I, X^I]_{lb} + \frac{1}{4}[X^I, X^J]_{lb}[X^I, X^J]_{lb}\right), \quad (34)$$

from which we see that the gauge field $\omega$ is a Lagrange multiplier for the constraint $K(\sigma) \equiv [P^I, X^I]_{lb} = 0$, where $K(\sigma)$ is the generator of SDiff. We have supposed in the above analysis that the membrane has the topology of a sphere. If it has some other topology, for which the first homology group is non-trivial, then one must include additional harmonic vectors in (27) which act as Lagrange multipliers for additional *global* constraints. For example, for a membrane of topology of $T^2$ a basis for the generators of SDiff can be found for which the Lie algebra is [16]

$$[L_\mathbf{m}, L_\mathbf{n}] = \mathbf{m} \times \mathbf{n} \ L_{\mathbf{m+n}}$$
$$[\mathbf{P}, L_\mathbf{m}] = \mathbf{m} \ L_\mathbf{m} \qquad [P_1, P_2] = 0. \qquad (35)$$

where $\mathbf{m} = (m_1, m_2)$, and $\mathbf{P} = (P_1, P_2)$ are the generators for the global constraints associated with the representatives of $H_1(T^2, \mathbf{R})$.

Finally let us return to the membrane in a three-dimensional spacetime. According to the above analysis this is equivalent to a dimensionally reduced *two-dimensional* Yang-Mills theory. The rather similar problem of two-dimensional Yang-Mills theory on a cylinder has been addressed recently by Rajeev [17] who has shown that it too can be reduced to a finite dimensional system. Perhaps there is a deeper connection here between two-dimensional Yang-Mills theory and three-dimensional Chern-Simons gauge theories.

## Acknowledgements

I am grateful to Eric Bergshoeff, Ergin Sezgin and Yoshiaki Tanii for the collaboration on ref. [5], on which part of this article is based. I also wish to thank Katsumi Itoh for bringing ref. [17] to my attention.

## References

[1] P.A.M. Dirac, *Proc. Roy. Soc.* **268** (1962)57.

[2] J. Hoppe, *Ph.D. Thesis*, M.I.T. 1982; Quantum theory of a relativistic surface, *Aachen preprint* PITHA-86/24.

[3] B. de Wit, J. Hoppe and H. Nicolai, On the quantum mechanics of supermembranes, *Karlsruhe preprint* KA-THEP 6/88.

[4] J. Hughes, J. Liu and J. Polchinski, *Phys. Lett.* **B180** (1986)370; E. Bergshoeff, E. Sezgin and P.K. Townsend, *Phys. Lett.* **B189** (1987)75; A. Achúcarro, J.M. Evans, P.K. Townsend and D.L. Wiltshire, *Phys. Lett.* **B198** (1987)441.

[5] E. Bergshoeff, E. Sezgin, Y. Tanii and P.K. Townsend, Super $p$-branes as gauge theories of volume preserving diffeomorphisms, *Trieste preprint*, 1988.

[6] P.K. Townsend, *Phys. Rev.* **D15** (1977)2802.

[7] P. Howe and R. Tucker, *J. Phys.* **A10** (1977)L155.

[8] T. Curtright, Extrinsic geometry of superimmersions, *CERN preprint* TH. 4924/87.

[9] A. Achúcarro and P.K. Townsend, *Phys. Lett.* **B180** (1986)89; A. Achúcarro, *M.Sc. Thesis*, Univ. of the Basque Country, Bilbao, Spain, 1986.

[10] M. Blencowe, *Class. Quantum Grav.* **6** (1989) 443.

[11] E. Witten, *Nucl. Phys.* **B311** (1988/89) 46.

[12] B. Biran, E.G. Floratos and G.K. Savvidy, *Phys. Lett.* **B198** (1987)329.

[13] K. Fujikawa, *Phys. Lett.* **B213** (1988) 425.

[14] E.Bergshoeff, M. Blencowe and K.S. Stelle, in preparation.

[15] V. Arnold, *Annales de l'Institut Fourier*, **XVI(1)** (1966) 319.

[16] E.G. Floratos and J. Illiopoulos, *Phys. Lett.* **B201** (1988)237.

[17] S.G. Rajeev, *Phys. Lett.* **B212** (1988)203.

DEPARTMENT OF APPLIED MATHEMATICS AND THEORETICAL PHYSICS, SILVER STREET, CAMBRIDGE CB3 9EW, ENGLAND.

# 8

## The Space of 2d Quantum Field Theories
*Emil J. Martinec*

Understanding of conformally invariant quantum field theories (cft's) in two dimensions has advanced rapidly in recent years (*cf.* the talks of E. Verlinde, L. Alvarez-Gaumé, and G. Segal). In the context of string theory, conformal theories are classical solutions of the string equations of motion[1]. Roughly speaking, this is because the string wave operator is the dilation operator of the 2d cft; the ground state wavefunction is in the kernel of this operator, yielding the linearized equation of motion $\Delta\Phi = 0$ (nonlinear effects can also be understood within the framework of the 2d theory). The analogue in 1d qft (quantum mechanics) is to think of the wave operator of linearized quantum field theory as the Hamiltonian for the quantum mechanical evolution in spacetime of single particle states [1]; $X$ is a quantum mechanical variable which describes the coordinate position of a particle in spacetime, and zero-energy states $\Phi(X)$ have $\Delta\Phi = 0$ and are thus solutions to the wave equation. Thus 2d qft provides an understanding of the quantum mechanics of single strings, where we now regard $X$ as a 2d quantum field. What we lack at the present time is a good notion of what the configuration space of string theory is. To borrow again from field theoretic concepts, in Yang-Mills theory configuration space is the space of connections modulo gauge transformations on a principal fiber bundle. The equations of motion select the subspace of configurations which are classical solutions of the theory. In string theory we are in the position of knowing a large variety of classical solutions without having a satisfactory configuration space in which to embed them. Configuration space is likely to be necessary in order to find a geometric understanding of the theory, to compute nonperturbatively, and so on.

One natural guess is that, if conformal quantum field theories are classical solutions, then the space of all quantum field theories is the configuration space, with conformal invariance the equation of mo-

---

[1] Actually, in string theory there is also a condition on the central extension of the Virasoro algebra. We will ignore this subtlety, as it is not a central issue in the present discussion.

tion. This motivates a study of nonconformal quantum field theory. Much less is known about the generic 2d quantum field theory, because we lose the power of the Virasoro algebra in analyzing its structure. Moreover, the amount of data we must supply about the 2d parameter space increases from merely conformal structure to metric structure. In these notes I would like to outline a few useful notions that survive the passage from conformal to generic qft, and paint a general picture of the space of 2d qft's quite similar to one familiar in ordinary spacetime field theory (*cf.* [2]). I will distinguish between 2d *parameter space* $\Sigma$ and the $D$ dimensional *spacetime* $\mathcal{M}$ in which the string evolves. For the most part our focus will be on the qft in 2d rather than the string theory in spacetime.

First let us define some basic objects. The action $S[\phi]$ of quantum field theory with field variable $\phi$ defines a probability distribution on function space $\{\phi\}$ (*e.g.* $\phi \in \text{Map}(\Sigma, \mathcal{M})$)

$$\mathcal{P}[\phi] = \frac{e^{-S[\phi]}}{\int \mathcal{D}\phi \, e^{-S[\phi]}}, \qquad (1)$$

where $\mathcal{D}\phi$ denotes a formal integration measure on $\{\phi\}$. *Correlation functions* are moments of $\mathcal{P}[\phi]$

$$\langle\langle f_1[\phi(z_1)] \cdots f_n[\phi(z_n)] \rangle\rangle \equiv \int \mathcal{D}\phi \, f_1[\phi(z_1)] \cdots f_n[\phi(z_n)] \mathcal{P}[\phi] \,. \qquad (2)$$

We may regard the theory as defined by its correlation functions, which are the set of all possible measurements in the qft. Note that the set of correlations may have several different representations (one example being boson-fermion equivalence in 2d free field theory). Typically, at long distances the randomness of $\phi$ washes out correlations

$$\langle\langle \phi(z_1)\phi(z_2) \rangle\rangle - \langle\langle \phi(z_1) \rangle\rangle \langle\langle \phi(z_2) \rangle\rangle \sim e^{-m|z_2-z_1|} |z_2 - z_1|^{-4h} \,. \qquad (3)$$

(Henceforth we write $\langle f_1 \cdots f_n \rangle$ to have the disconnected parts of the correlations removed.) Long distance is defined by $|z_2 - z_1| \gg \frac{1}{m}$, where $\frac{1}{m}$ is called the *correlation length*.

*Conformal theories* have the property of infinite correlation length: $m = 0$. Then the leading exponential behavior disappears, and correlations survive to long distances, decreasing only with a power of distance. Correlation functions, depending on distances $d(z_1, z_2)$ ($= |z_2 - z_1|$ here), are intrinsically metric dependent (although in the special case of conformal theories they depend only on conformal structure, ignoring the central extension of the conformal algebra). Therefore define the canonical object $T_{ab}$, the *stress tensor*, by

$$\frac{\delta}{\delta g^{ab}(w)} \langle f_1(z_1) \cdots f_n(z_n) \rangle = \langle T_{ab}(w) f_1(z_1) \cdots f_n(z_n) \rangle \,. \qquad (4)$$

## 8. 2d Quantum Field Theories

Coordinate independence requires

$$\int_\Sigma \delta_v g^{ab} \frac{\delta}{\delta g^{ab}} \langle f_1 \cdots f_n \rangle = \sum_{i=1}^n \langle f_1 \cdots \delta_v f_i \cdots f_n \rangle, \quad (5)$$

$\forall\, \delta_v g^{ab} = \nabla^{(a} v^{b)}$. In other words, the correlation functions carry a natural action of the algebra of vector fields. Similarly, $g^{ab}\frac{\delta}{\delta g^{ab}}$ is the generator of local scale transformations. Conformal invariance is the statement

$$g^{ab} T_{ab} = \frac{c}{48\pi} R^{(2)} \quad (6)$$

in all correlations functions; $c$ is the central extension of the Virasoro algebra, and $R^{(2)}$ is the local scalar curvature. More precisely, we should say that $\langle g^{ab} T_{ab} \rangle = 0$ away from sources $f_i$;[2] at the sources we get contributions from the scaling properties of the sources.

Zamolodchikov [3] introduced some fundamental structures in 2d qft, including a quantity whose restriction to scale invariant theories is the central term $c$. Let's work in local complex coordinate charts; then $g^{ab} T_{ab} \equiv T_{z\bar{z}} \equiv -\Theta$, and the traceless components are $T_{zz} \equiv T$, $T_{\bar{z}\bar{z}} \equiv \overline{T}$. Now construct the two-point correlation functions (take $\Sigma = \mathbf{C}$ for now)

$$\begin{aligned} \langle T(z) T(0) \rangle &\equiv F(t)/z^4 \\ \langle T(z) \Theta(0) \rangle &\equiv H(t)/(z^3 \bar{z}) \\ \langle \Theta(z) \Theta(0) \rangle &\equiv G(t)/(z\bar{z})^2 \\ t &\equiv \log(z\bar{z})\,. \end{aligned} \quad (7)$$

One can think of $t$ as defining a one-parameter family of metrics $g_{ab}(t) = e^{2t} g_{ab}(0)$. The conservation of stress-energy (5) $\bar{\partial} T = \partial \Theta$, $\bar{\partial}\Theta = \partial T$ implies

$$\begin{aligned} \partial_t F &= \partial_t H - 3H \\ \partial_t H - 3H &= \partial_t G - 2G\,. \end{aligned} \quad (8)$$

Define the quantity $c(t) = 2F + 4H - 6G$ [3]; then

$$\partial_t c = -12 G \geq 0 \quad (9)$$

and $\partial_t c = 0$ only when $\Theta = 0$, i.e. when the theory is conformally invariant.

To exploit this result, we need to define what we might call the space of quantum field theories (or at least a neighborhood of it about the generic theory under discussion). We will then see that $c$ is a Morse

---

[2] In other words $R^{(2)}(w)$ in (6) contributes only to the disconnected parts $\langle\langle R^{(2)}(w)\rangle\rangle \langle\langle f_1(z_1) \cdots f_n(z_n)\rangle\rangle$ unless $w = z_i$ for some $i = 1, ..., n$.

function on this space, whose critical points are the conformal theories – the classical solutions. In spacetime field theory, the analogue of $c$ is the action; for instance, the action in Euclidean Yang-Mills field theory has as critical points the self-dual Yang-Mills fields, which are the solutions of the Euclidean equations of motion.

Choose a basis of quantum fields $\mathcal{O}_i(z)$ (which we denoted $f_i[\phi]$ above). The choice of basis depends on the theory. We may expand

$$S(g) = S_0 + \int_\Sigma \lambda^i(g)\mathcal{O}_i(g,z) \qquad (10)$$

about a reference theory $S_0$, in such a way that the field $\mathcal{O}_i(g,z) = \frac{\partial}{\partial g^i}S(g,z)$ is the derivative of the local action density. There is a great deal of freedom in this procedure depending on the choice of basis $\mathcal{O}_i(g,z) \sim \mathcal{O}_i(z) + o(g)$, the functions $\lambda^i(g) \sim g^i + o(g^2)$, the metric scale $t$, and so on; fixing choices is called choosing a *renormalization scheme*. The *coupling constants* $g^i$ are coordinates that parametrize a family of qft's in the neighborhood of the reference theory with measure defined by $S_0 = S(g=0)$. Acting inside the functional integral, the $\mathcal{O}_i$ are tangent vectors to the space of theories. Under infinitesimal scale transformations

$$S(g,z) \to S(g,z) - dt\Theta(g,z) \; ; \qquad (11)$$

one can also expand

$$\Theta = \sum_i \beta^i(g)\mathcal{O}_i(g,z) \implies \frac{dg^i}{dt} = \beta^i(g) \qquad (12)$$

*i.e.*, $\Theta$ defines a vector field ($\beta^i(g)$ in coordinates) giving a one-parameter flow in the space of quantum field theories under rescaling of the 2d metric. That is, the rescaling operator $\Theta$ relates a theory with coupling constants $g^i$ and fiducial length scale $e^t$ (some number of correlation lengths) to another with couplings $g^i + \beta^i dt$ at scale $e^{t+dt}$. Thus implicitly $c(t) = c(g,t)$ is a function of couplings in our coordinate patch of theory space; we may regard $c(g) = c(g,0)$ as the advertised Morse function on the space of qft's. An important difference from conformal theories is the need to specify a reference length scale for measuring correlations; this scale drops out of the theory at a critical point, where correlations depend only on ratios of distances (moduli). The condition for scale invariance is $\beta^i = 0$; the couplings, and hence the theory, are invariant under rescaling. Note that $\langle \mathcal{O}_i(1)\mathcal{O}_j(0)\rangle_g \equiv G_{ij}(g)$ is the inner product of coordinate tangent vectors, and defines a metric on theory space [3].

## 8. 2d Quantum Field Theories

We might worry that the perturbations that we add to deform the theory cause such violent changes in the measure that the perturbed theory does not exist. Let's investigate how this could occur. Suppose $S_0$ describes a scale invariant qft (a cft). Then the natural basis for $\mathcal{O}_i$ diagonalizes the scaling operator linearized at $S_0$, i.e. $g = 0$, with eigenvalues $h_i$. Noting that

$$\mathcal{O}_i(e^t z) d^2(e^t z) \sim e^{-2(h_i-1)t} \cdot \mathcal{O}_i(z) d^2 z , \tag{13}$$

we have the following three possibilities:

$$\begin{array}{ll} 1)\ h_i > 1 & \text{irrelevant} \\ 2)\ h_i < 1 & \text{relevant} \\ 3)\ h_i = 1 & \text{marginal} . \end{array} \tag{14}$$

As for (1), such perturbations drastically alter the local (high momentum) behavior, since they grow at short distances; we have no right to describe the resulting theory as a small perturbation of $S_0$. Perturbations of class (2) are fine – they don't change short distances, but smoothly change the long distance properties of the measure. For case (3) it depends on the nonlinear terms in $\beta^i$, since there are potentially logarithmic corrections to scaling in this case. It can happen (see below) that there is a subspace of $g$'s for which $\beta^i(g) = 0$, leading to a *moduli space* of conformal field theories. The family of theories defined by perturbations of classes (2) and (3) are the *renormalizable* qft's.

One can calculate in the neighborhood of a given soluble conformal field theory in an expansion in $g$,

$$\begin{aligned} \langle \mathcal{O}_1 \cdots \mathcal{O}_n \rangle_g &= \left\langle \mathcal{O}_1 \cdots \mathcal{O}_n e^{-\lambda^i \int \mathcal{O}_i} \right\rangle_0 \\ &= \sum_n \frac{1}{n!} \left\langle \mathcal{O}_1 \cdots \mathcal{O}_n (-\lambda^i \int \mathcal{O}_i)^n \right\rangle_0 . \end{aligned} \tag{15}$$

The neighborhood of the space of field theories about $S_0$ is described by certain integrated correlation functions in the reference theory. Also,

$$\langle \mathcal{O}_i(1) \mathcal{O}_j(0) \rangle = G_{ij}(g) \tag{16}$$

is a positive definite metric on theory space [3] (for unitary qft's, i.e. those that admit a probability interpretation). The positivity of $G$ is responsible for the monotonicity of $\partial_t c = -12(\beta, \beta)$. Using the equation for scaling of correlation functions

$$(\partial_t + \beta^i \frac{\partial}{\partial g^i} + \sum_{a=1}^n \Gamma_a) \langle \mathcal{O}_1 \cdots \mathcal{O}_n \rangle = 0 , \tag{17}$$

where $\Gamma_a$ computes the effect of rescaling on the field $\mathcal{O}_a$, we can compute $\beta^i(g)$ in Taylor series about $g^i = 0$ from, e.g., the scaling behavior of the two-point correlation function. The result is

$$\beta^i(g) = 2(1 - h_i)g^i + C^i_{jk}g^j g^k + D^i_{jkl}g^j g^k g^l + \ldots \tag{18}$$

Perhaps the simplest set of conformal field theories is the so-called discrete series: If one looks for unitary representations of the Virasoro algebra with central charge $c < 1$, one is restricted to the list of values [4]

$$c_m = 1 - \frac{6}{m(m+1)}, \qquad m = 2, 3, \ldots \tag{19}$$

The relevant operators have scale dimensions

$$h_p = \frac{p^2 - 1}{4m(m+1)} \qquad p = 1, \ldots, 2m - 1 \quad p \neq m, m+1 \,. \tag{20}$$

The operator algebra and discrete symmetries, as well as arguments based on integrable lattice statistical mechanical models, identify these theories as the fixed points of the set of actions [5,6]

$$S^{(m)} = \int \partial \phi \bar{\partial} \phi + g \phi^{2(m-1)} \tag{21}$$

with

$$\begin{aligned} \mathcal{O}_p &= \phi^{p-1} & p = 1, \ldots, m-1 \\ &= \phi^{p-3} & p = m+2, \ldots, 2m-1 \end{aligned} \tag{22}$$

i.e., all powers of $\phi$ up to $\phi^{2(m-2)}$; $\phi^{2m-3}$ is related to $\partial \bar{\partial} \phi$ by the equation of motion. Weak coupling ($g \sim \frac{1}{m}$, $m \gg 1$) fixed points have $\partial \phi \bar{\partial} \phi$, $\phi^{2(m-1)}$ both approximately marginal. Perturbations such as $g' \int \phi^{2(m-2)}$ violate linearized scale invariance only slightly, and one finds a self-consistent expansion in powers of $g'$

$$\beta_{2(m-2)}(g') = \tfrac{4}{m}g' - \tfrac{4}{\sqrt{3}}(g')^2 + o((g')^3) \,. \tag{23}$$

There is a new fixed point $\beta = 0$ at $g'_* = \frac{\sqrt{3}}{m} + o(\frac{1}{m^2})$ [3,6]. We find that [3,6] $S^{(m)}$ flows to $S^{(m-1)}$; of course the theory must belong to the list (19). This is verified by integrating $\partial_t c = -12\beta^2$:

$$\begin{aligned} c(g'_*) - c(0) &= -12 \int_0^{g'_*} \beta(g') dg' = -\frac{12}{m^3} + o(\frac{1}{m^4}) \\ &= c_{m-1} - c_m \,. \end{aligned} \tag{24}$$

The tangent space of relevant operators drops two dimensions each

## 8. 2d Quantum Field Theories

time we flow $m \to m - 1$, i.e. $g_*$ is a saddle point with two stable directions, $2(m-3)$ unstable directions when considered in the context of the $m^{th}$ theory. The generic perturbation mixes all relevant fields $\mathcal{O}_i$, e.g. each $\beta^i$ will have a term $C^i(g')^2$ at second order which destabilizes all the $g^i$; the particular perturbation above is somewhat special in that it only involved one scaling field – $C^i = 0$ except for the one in $\beta_{g'}$. Thus we see that there is a rich structure of fixed points and flows in the space of field theories for small $c$.

The $c < 1$ theories have no moduli; there are no marginal operators. The simplest example of a moduli space of conformal theories occurs at $c = 1$. There are several cases [7]. First consider strings on a circle $\mathcal{M} = S^1$ of radius $R$. The action is $S = \frac{1}{2\pi R^2} \int \partial\phi \bar{\partial}\phi$, where $\phi \equiv \phi + 2\pi$. The spectrum of scale dimensions is

$$h_{pq} = \tfrac{1}{2}(\tfrac{p}{2R} + Rq)^2 + \text{positive integer}, \qquad p, q \in \mathbf{Z} \qquad (25)$$

There is a marginal direction $\mathcal{O} = \partial\phi\bar{\partial}\phi$ since we have a solution for all radii $R$. Note also that the spectra are identical for $R$ and $\frac{1}{2R}$; in fact one can show that not only the spectra are the same but in fact the theories are isomorphic. So our first ingredient of the $c = 1$ moduli space is the half-line $R \geq \frac{1}{\sqrt{2}}$. In addition, if a space $\mathcal{M}$ carries a discrete symmetry $\Gamma$, it makes perfect sense to consider the qft on $\widetilde{\mathcal{M}} = \mathcal{M}/\Gamma$ [8,7]. Here we apply the method to $\mathcal{M} = S^1$, $\Gamma = \mathbf{Z}/2\mathbf{Z}$ acting as $\phi \to -\phi$. The allowed scale dimensions are (25) together with $h = \frac{1}{16} + \frac{1}{2}(\text{pos. int.})$. It turns out that the $\widetilde{\mathcal{M}}(R = 1)$ theory is identical to the $\mathcal{M}(R = \sqrt{2})$ theory. At $R = \sqrt{2}$ the fields with $p = \pm 4$, $q = 0$ or $p = 0$, $q = \pm 1$ are marginal; what is not obvious is that $\beta = 0$ for a perturbation by these fields. So in fact the line of orbifold theories meets the circle line (and ends) at this point. Finally, note that the cft at the self-dual point $\mathcal{M}(R = \frac{1}{\sqrt{2}})$ has extra $h = 1$ fields. Generically the Hilbert space of $c = 1$ is a representation of the (centrally extended) loop group on $U(1)$; at the special point $R = \frac{1}{\sqrt{2}}$ the theory realizes the loop group on $SU(2)$ [9], and there are new marginal operators made out of the currents $j$ ($p = \pm 2$, $q = 0$ or viceversa). Then one can quotient the self-dual theory by any discrete subgroup of $SU(2)$, and find three new theories corresponding to the regular solid groups $\Gamma = \mathcal{T}$, $\mathcal{O}$, and $\mathcal{I}$. Thus the full moduli space is as follows [7]:

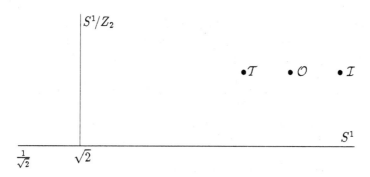

In known examples the non-manifold parts of the moduli space of conformal theories correspond to theories where a new marginal direction appears, such that $\beta = 0$. Seiberg [10] conjectures that this is the only way that singularities in the moduli space arise. Seiberg has studied the moduli space of cft's where $\mathcal{M}$ is Ricci-flat Kähler by analysis of superstring theory on $\mathbf{R}^{10-\dim\mathcal{M}} \times \mathcal{M}$. Symmetries induced by covariantly constant spinors (supersymmetries) on these spaces severely restrict the structure of the metric on the cft moduli space (which is none other than $G_{ij}(g)$). In particular he finds that the local structure of the K3 moduli space is the homogeneous space $\frac{O(20,4)}{O(20)\times O(4)}$. Further work has elaborated the structure of cft moduli spaces [11].

To summarize, we can grade the space of qft's by $c$; we have exhibited the space of cft's with $c \leq 1$ (the mentioned $c = 1$ theories are believed to exhaust the list). The characteristic number $c$ can be regarded as the number of degrees of freedom at asymptotically large 2d distances, measured in units of free scalar fields $c = 1$. The $c < 1$ critical points are isolated; $c = 1$ models generically have moduli. It is hoped that enough can be understood about 2d qft to map out the configuration space.

The description of conformal theories as the fixed points of scalar field theories has a straightforward extension to supersymmetric 2d theories [12]; a beautiful aspect of the set of ideas I have outlined here is the case of N=2 (*i.e.* complex) 2d supersymmetry. In this case the polynomial potential theories appear to describe spacetimes which are algebraic varieties [12,13], where the polynomial ideal is generated by the superpotential!

## Acknowledgements

This work was supported in part by DOE grants DE-FG02-84ER-45144 and NSF grant NSF-PHY-86-57788.

# References

[1] B.S. DeWitt, Dynamical Theory in Curved Spaces. I. A Review of the Classical and Quantum Action Principals, *Rev. Mod. Phys.* **29** (1957) 377.

[2] M. Atiyah and R. Bott, The Yang-Mills Equations Over Riemann Surfaces, *Phil. Trans. R. Soc. London* **A308** (1982) 523.

[3] A.B. Zamolodchikov, Renormalization Group and Perturbation Theory About Fixed Points in Two-Dimensional Field Theory, *Sov. J. Nucl. Phys.* **46** (1988) 1090.

[4] D. Friedan, Z. Qiu, and S.H. Shenker, Conformal Invariance, Unitarity, and Two-Dimensional Critical Exponents, in *Vertex Operators in Mathematics and Physics*, ed. J. Lepowsky et. al., Springer-Verlag, 1984.

[5] A.B. Zamolodchikov, Conformal Symmetry and Multicritical Points in Two-Dimensional Quantum Field Theory, *Sov. J. Nucl. Phys.* **44** (1986) 529. See also D. Huse, Exact Exponents for Infinitely Many New Multicritical Points, *Phys. Rev.* **B30** (1984) 3908.

[6] J. Cardy and A. Ludwig, Perturbative Evaluation of the Conformal Anomaly at New Critical Points with Applications to Random Systems, *Nucl. Phys.* **B285** (1987) 687.

[7] G. Harris, $SU(2)$ Current Algebra Orbifolds of the Gaussian Model, *Nucl. Phys.* **B300** (1988) 588; P. Ginsparg, Curiosities at $c = 1$, *Nucl. Phys.* **B295** (1988) 153.

[8] L. Dixon, J. Harvey, C. Vafa, and E. Witten, Strings on Orbifolds, *Nucl. Phys.* **B261** (1985) 620; **B274** (1986) 285.

[9] G. Segal, Unitary Representations of Some Infinite-Dimensional Groups, *Comm. Math. Phys.* **80** (1981) 301; I. Frenkel and V. Kac, Basic Representations of Affine Lie Algebras and Dual Resonance Models, *Inv. Math.* **62** (1980) 23.

[10] N. Seiberg, Observations on the Moduli Space of Superconformal Field Theories, *Nucl. Phys.* **B303** (1988) 286.

[11] S. Cecotti, S. Ferrara, and L. Girardello, Geometry of Type II Superstrings and the Moduli of Superconformal Field Theories, CERN-TH-5080, June 1988; D. Kutasov, Geometry on the Space of Conformal Field Theories and Contact Terms, Weizmann preprint WIS-88/55/Oct-PH.

[12] D. Kastor, E. Martinec, and S.H. Shenker, *Nucl. Phys.* **B316** (1989) 590.

[13] E. Martinec, Algebraic Geometry and Effective Lagrangians, Chicago preprint EFI-88-76 (November, 1988).

THE ENRICO FERMI INSTITUTE AND DEPARTMENT OF PHYSICS, UNIVERSITY OF CHICAGO, CHICAGO, ILLINOIS 60637.

# 9

# Chern–Simons Forms and Cyclic Cohomology

*D. Quillen*

## 1. Introduction

In [1] cyclic cohomology appears as a noncommutative analogue of de Rham cohomology suggested by $K$-theory and index theory in noncommutative geometry. It is closely tied to $K$-theory on one hand and to differential forms on the other. It is therefore natural to study cyclic cohomology using known methods linking $K$-theory and differential forms such as the Chern-Weil approach to characteristic classes of vector bundles based on connections and curvature.

In [4] a theory of cochains on an algebra $A$ is developed, where cyclic cocycles are constructed as analogues of Chern character and Chern-Simons forms over a differential graded algebra made of cochains. These cocycles are important in the study of the cyclic cohomology of an extension of the algebra [5].

In the present paper we consider Chern-Simons forms over the bigraded differential algebra of cochains having values in the noncommutative differential forms $\Omega_A$ over $A$. The DG algebra $\Omega_A$ plays an important role in Connes' approach to cyclic cohomology, because cyclic cocycles are equivalent to closed traces on $\Omega_A$. The Chern-Simons forms provide the link between Connes' theory and the algebra cochain theory.

The organization of this paper is as follows. In §1 we study Chern-Simons forms in a general noncommutative differential graded algebra. The Chern-Simons forms of different degree are linked by certain relations, called $S$-relations, because they are connected with the $S$-operation in cyclic cohomology. Adapting an idea of Weil, we introduce the universal DG algebra generated by a 'connection' form. This noncommutative Weil algebra has a natural bigrading suggested by cyclic theory, and we determine various cohomology groups associated to the bigrading in analogy with the BRS algebras of [2].

In §2 we begin with a fairly down to earth account of the theory of algebra cochains. We then discuss the Chern-Simons forms over the bigraded algebra of cochains with values in $\Omega_A$, and the cyclic cocycles arising from them by applying a closed trace on $\Omega_A$. We show these cocycles coincide with the iterates of Connes $S$-operation applied to the cyclic cocycle corresponding to the closed trace.

## 2. Chern-Simons forms and Weil algebra

### 2.1 Chern character and Chern-Simons forms

Let $R = \{R^n\}$ be a differential graded (DG) algebra over the complex numbers $\mathbf{C}$ with differential $d : R^n \to R^{n+1}$. We do not assume it has an identity.

Let $[R, R]$ be the subspace spanned by commutators $[x, y] = xy - (-1)^{|x||y|} yx$, where $|x|$ denotes the degree of $x$. By a trace on $R$ we will mean a degree zero map $\tau$ from $R$ to a complex $V$ such that $d\tau = \tau d$ and $\tau[R, R] = 0$.

For example, if $V = \mathbf{C}[p]$ is the complex which is $\mathbf{C}$ in degree $p$ and zero elsewhere, then a trace $R \to V$ is equivalent to a linear functional on $R^p$ which vanishes on $dR^{p-1} + [R, R]^p$. Such a linear functional is a kind of noncommutative analogue of a closed current [1], II.

As another example, let $\natural : R \to R_\natural$ be the canonical surjection, where $R_\natural = R/[R, R]$ is the commutator quotient complex. Then $\natural$ is the universal trace on $R$ in the sense that any other trace $\tau : R \to V$ is induced from $\natural$ by a unique map of complexes $R_\natural \to V$.

Let $\tau$ be a trace on $R$ and let $A$ be an element of $R$ of degree one. Viewing $A$ as analogous to a connection form, we let $F = dA + A^2 \in R^2$ be its curvature, and we construct the Chern character form $\tau(F^n/n!)$. From the Bianchi identity $(d + \mathrm{ad}A)F = dF + [A, F] = 0$, we derive in the usual way

$$d\{\tau(F^n)\} = \tau(d(F^n)) = \tau(-[A, F^n]) = 0 \qquad (1)$$

that the Chern character form is closed.

By refining this argument, one shows that if $A_t$ is a family of elements of $R^1$ depending in a polynomial fashion on $t$, then the following infinitesimal homotopy formula holds

$$\partial_t\{\tau(F_t^n/n!)\} = d\{\tau(\dot{A}_t F_t^{n-1}/(n-1)!)\} \qquad (2)$$

where $\dot{A}_t = \partial_t A_t$ and $F_t = dA_t + A_t^2$. In the case $A_t = tA$ this may be integrated to obtain the transgression formula

$$\tau(F^n/n!) = d\{\mathrm{cs}_{2n-1}(A, \tau)\} \qquad (3)$$

where $\mathrm{cs}_{2n-1}(A, \tau)$ is the Chern-Simons form

$$\mathrm{cs}_{2n-1}(A, \tau) = \int_0^1 \tau(A F_t^{n-1}/(n-1)!) dt \qquad (4)$$

## 9. Chern–Simons Forms and Cyclic Cohomology

We consider a geometric example of these constructions. Let $P$ be a principal bundle over the manifold $M$ with group $U(r)$, let $\Omega(P)$ be the DG algebra of differential forms on $P$, $R = M_r(\Omega(P))$ the matrix differential forms, and let $\tau : R \to \Omega(P)$ be the usual matrix trace. We suppose given a connection in $P$. Its connection and curvature forms $A$, $F$ can be viewed as elements $R$ of degree 1 and 2 such that $dA + A^2 = F$. The expressions (3), (4) are differential forms on $P$ in this case, and one knows that the Chern character form comes from a differential form on the base $M$. Hence for $2n > \dim(M)$ the Chern-Simons form is closed, and by applying this argument over $M \times \mathbf{R}$, one shows that its cohomology class is independent of the connection for $2n > \dim(M) + 1$.

### 2.2 Periodic complex for $R$

The Chern character and the Chern-Simons forms of different degrees are linked by certain relations, which we call $S$-relations because they are connected with the $S$-operation on cyclic cohomology. The $S$-operation raises degree by two and is apparently related to Bott periodicity in some mysterious way. We now discuss some tools needed to formulate the $S$-relations.

Let $\Omega_R^1$ be the noncommutative 1-forms over $R$ [4]. It is a DG bimodule over $R$, that is, a complex with left and right multiplication by $R$ compatible with differentials. There is a canonical map of complexes $\partial : R \to \Omega_R^1$, which is a derivation: $\partial(xy) = \partial x\, y + x\, \partial y$ having a universal mapping property relative to derivations from $R$ to bimodules over $R$. If $\tilde{R} = \mathbf{C}1 \oplus R$ denotes the unital DG algebra obtained by adjoining an identity, we have an isomorphism

$$\Omega_R^1 \cong R \otimes \tilde{R} \qquad \partial x\, \tilde{r} \longleftrightarrow x \otimes \tilde{r} \qquad (5)$$

for $x \in R$, $\tilde{r} \in \tilde{R}$. This description of $\Omega_R^1$ may be used to construct it concretely as $R \otimes \tilde{R}$ with suitable left and right multiplication cite[p. 99]1

Let $\Omega_{R,\natural}^1 = \Omega_R^1/[R, \Omega_R^1]$ and let $\natural : \Omega_R^1 \to \Omega_{R,\natural}^1$ be the canonical surjection. We have

$$\begin{aligned} \natural(\partial(xy)\tilde{r}) &= \natural(\partial x\, y\tilde{r}) + \natural(x\partial y\, \tilde{r}) \\ &= \natural(\partial x\, y\tilde{r}) + (-1)^{|x|(|y|+|\tilde{r}|)}\natural(\partial y\, \tilde{r}x) \end{aligned} \qquad (6)$$

Hence if $u : \Omega_{R,\natural}^1 \to V$ is a map of complexes and we set

$$\psi(x, \tilde{r}) = u\natural(\partial x\, \tilde{r}) \qquad (7)$$

then $\psi : R \times \tilde{R} \to V$ is a bilinear map of degree zero compatible with differentials:

$$d\psi(x, \tilde{r}) = \psi(dx, \tilde{r}) + (-1)^{|x|}\psi(x, d\tilde{r}) \qquad (8)$$

satisfying the cocycle condition

$$\psi(xy,\tilde{r}) = \psi(x,y\tilde{r}) + (-1)^{|x|(|y|+|\tilde{r}|)}\psi(y,\tilde{r}x) \tag{9}$$

Conversely given such a $\psi$, we use (7) to define a map $\Omega^1_R \to V$ such that $\partial x\,\tilde{r} \mapsto \psi(x,\tilde{r})$. This is a map of complexes by (8) and it vanishes on $[R,\Omega^1_R]$ by (9), so we obtain such a map $u$.

We therefore have an equivalence between maps of complexes $\Omega^1_{R,\natural} \to V$ and bilinear maps $\psi : R \times \tilde{R} \to V$ satisfying (8), (9).

If we had used the isomorphism similar to (5) but with $\tilde{R}$ on the right, then (9) would have taken the familiar form $b\psi = 0$ [1]. The above notation is better suited for the discussion of cyclic cohomology below.

There are canonical maps of complexes $\bar{\partial} : R \to \Omega^1_{R,\natural}$ and $\beta : \Omega^1_{R,\natural} \to R$ given by

$$\bar{\partial}x = \natural(\partial x) \qquad \beta\{\natural(\partial x\,\tilde{r})\} = -[x,\tilde{r}] \tag{10}$$

$\beta$ corresponds to the bilinear map $\psi(x,\tilde{r}) = -[x,\tilde{r}]$. One has $\beta\bar{\partial} = \bar{\partial}\beta = 0$, so we obtain a periodic complex

$$\xrightarrow{\bar{\partial}} \Omega^1_{R,\natural} \xrightarrow{\beta} R \xrightarrow{\bar{\partial}} \Omega^1_{R,\natural} \xrightarrow{\beta} \tag{11}$$

consisting of complexes. It can be viewed as a double complex periodic in the horizontal direction, provided signs are introduced to make the horizontal and vertical differentials anticommute.

The cokernel of $\beta$ is $R_\natural$, so (11) is obtained by splicing together copies of the sequence

$$0 \longrightarrow R_\natural \xrightarrow{\bar{\partial}} \Omega^1_{R,\natural} \xrightarrow{\beta} R \xrightarrow{\natural} R_\natural \longrightarrow 0 \tag{12}$$

This sequence is exact only at the right in general. However when $R$ is a free DG algebra, it is exact everywhere and we get an operation $S$ of degree 2 on the cohomology of $R_\natural$ as follows. Given $\alpha \in H^n(R_\natural)$, we represent it by a cocycle $t$ and use exactness to solve successively the equations

$$\natural(u) = t, \quad \beta(v) = du, \quad \bar{\partial}(w) = dv \tag{13}$$

Then $S\alpha \in H^{n+2}(R_\natural)$ is represented by the class of $w$. We shall see that a variant of this construction leads to the $S$-operation on cyclic cohomology.

## 2.3 S-relations

We begin with the following relations satisfied by the elements $(F^n/n!)$ of $R$ and $\natural(\partial A\, F^n/n!)$ of $\Omega^1_{R,\natural}$:

$$\begin{aligned} d(F^n/n!) &= \beta\{\natural(\partial A\, F^n/n!)\} \\ d\{\natural(\partial A\, F^n/n!)\} &= \bar\partial\{F^{n+1}/(n+1)!\} \end{aligned} \qquad (14)$$

Proof. $d(F^n) = -[A, F^n] = \beta\{\natural(\partial A\, F^n)\}$. $d\{\natural(\partial A\, F^n)\} = \natural\{(d+\operatorname{ad}A)(\partial A\, F^n)\} = \natural\{(d\partial A + A\partial A + \partial A\, A)F^n\} = \natural(\partial F\, F^n) = \bar\partial\{F^{n+1}/(n+1)\}$. □

The formulas (14) say that the $S$-operation on $H^{\cdot}(R_\natural)$, which is defined by the diagram chasing (13), takes the class of the Chern character form $\natural(F^n/n!)$ to the class of the next one. Of course this is not very interesting, since these classes are zero.

Consider next a family $A_t$ and set

$$\mu_n(A_t, \dot A_t) = \sum_{i=1}^n F_t^{i-1} \dot A_t F_t^{n-i} \qquad (15)$$

We note that this is an element of $R$ such that

$$\natural(\mu_n(A_t, \dot A_t/n!)) = \natural\{\dot A_t F_t^{n-1}/(n-1)!\} \qquad (16)$$

We have the following infinitesimal homotopy version of (14), where we write $\mu_{nt}$ for $\mu_n(A_t, \dot A_t)$

$$\begin{aligned} \partial_t(F_t^n) &= d\{\mu_{nt}\} - \beta\{\natural(\partial A_t\, \mu_{nt})\} \\ \partial_t\{\natural(\partial A_t\, F_t^n)\} &= -d\{\natural(\partial A_t \mu_{nt})\} + \bar\partial\{\mu_{n+1,t}/(n+1)\} \end{aligned} \qquad (17)$$

Proof. To simplify the notation we replace $A_t$, $F_t$, $\mu_{nt}$ by $A$, $F$, $\mu_n$. Using $\partial_t F = (d + \operatorname{ad}A)(\dot A)$, we have

$$\begin{aligned} \partial_t(F^n) &= \sum_1^n F^{i-l}(d + \operatorname{ad}A)(\dot A)F^{n-i} = (d + \operatorname{ad}A)(\mu_n) \\ &= d\mu_n + [A, \mu_n] = d\mu_n - \beta\{\natural(\partial A\, \mu_n)\} \end{aligned}$$

proving the first formula. Next we have

$$\begin{aligned} \partial_t\{\natural(\partial A\, F^n)\} &= \natural(\partial \dot A\, F^n + \partial A(d + \operatorname{ad}A)\mu_n) \\ &= \natural\{\partial \dot A\, F^n - d(\partial A\, \mu_n) + (\partial(dA) + [A, \partial A])\mu_n\} \\ &= -d\{\natural(\partial A\, \mu_n)\} + \natural(\partial \dot A\, F^n + \partial F\, \mu_n) \end{aligned}$$

Because $\bar{\partial}$ kills $[R, R]$ we have

$$\bar{\partial}(\mu_{n+1}/(n+1)) = \bar{\partial}(\dot{A}F^n) =$$

$$\natural\{\partial \dot{A}\, F^n + \dot{A}\sum_1^n F^{i-1}\partial F\, F^{n-i}\} = \natural(\partial \dot{A}\, F^n + \partial F\, \mu_n)$$

proving the second formula. □

Consider now the case $A_t = tA$, where $\dot{A}_t = A$, $F_t = tdA + t^2 A^2 = tF + (t^2 - t)A^2$, and put

$$\begin{aligned}\varphi_{2n-1} &= \int_0^1 \mu_n(tA, A)\, dt/n! \\ \psi_{2n} &= \int_0^1 \natural(\partial A\, \mu_n(tA, A))\, t dt/n!\end{aligned} \quad (18)$$

Integrating (16) yields

$$\natural(\varphi_{2n-1}) = \operatorname{cs}_{2n-1}(A, \natural) \quad (19)$$

in $R_\natural$, and integrating the relations (17) gives

$$\begin{aligned}F^n/n! &= d\varphi_{2n-1} - \beta\psi_{2n} \\ \natural(\partial A\, F^n/n!) &= -d\psi_{2n} + \bar{\partial}\varphi_{2n+1}\end{aligned} \quad (20)$$

Suppose now that $F^m = 0$, so that the Chern-Simons forms $\operatorname{cs}_{2n-1}(A, \natural)$, $n \geq m$ are closed. Then (19), (20) imply the relations

$$S[\operatorname{cs}_{2n-1}(A, \natural)] = [\operatorname{cs}_{2n+1}(A, \natural)] \quad (21)$$

for their cohomology classes.

In the situation of a principal $U(r)$ bundle $P$ mentioned above, we therefore have a link between the different Chern-Simons classes in the cohomology of $P$, which is defined in a strange algebraic way. It would be interesting to give this $S$-operation a geometric definition, hopefully connected with Bott periodicity, but this problem we have to leave to the future.

### 2.4 The Weil algebra and its bigrading

We now adopt an idea of Weil and introduce the universal DG algebra over which the above constructions can be done. We define the Weil algebra $W$ to be the DG algebra freely generated by a distinguished element $A$ of degree one. Thus given an element of degree one in a DG algebra $R$ there is a unique map of DG algebras $W \to R$ carrying $A$ to this element.

## 9. Chern–Simons Forms and Cyclic Cohomology

The Weil algebra can be described as the tensor algebra $\bar{C}\langle A, dA\rangle$ (noncommutative polynomial algebra without identity) with the generators $A$, $dA$, where the differential $d$ is defined in the obvious way. It can also be described as the tensor algebra $\bar{C}\langle A, F\rangle$, where $F$ has degree two, with differential defined to be the (anti)-derivation such that $dA = F - A^2$ and $dF = -[A, F]$.

The applications to cyclic cohomology we will consider as well as the theory of connections on holomorphic vector bundles suggest that the Weil algebra has the structure of a bigraded differential algebra where the universal 'connection' $A$ has degree $(0, 1)$ and the universal 'curvature' $F$ has degree $(1, 1)$. In order to establish this, as well as to introduce a convenient notation to handle this bigrading, let us consider the tensor algebra $W' = \bar{C}\langle \theta, d\theta\rangle$ with bigrading defined by requiring $\theta$ and $d\theta$ to have degrees $(0, 1)$ and $(1, 1)$ respectively. Let $d$ and $\delta$ be the derivations of $W'$ such that $d(\theta) = d\theta$, $d(d\theta) = 0$, $\delta(\theta) = -\theta^2$, $\delta(d\theta) = -[\theta, d\theta]$. Then $d$ and $\delta$ have degrees $(1, 0)$ and $(0, 1)$ respectively, and we have $d^2 = \delta^2 = d\delta + \delta d = 0$, since these are derivations vanishing on the generators. In this way $W'$ becomes a bigraded differential algebra.

In $W'$ we have $(d+\delta)\theta + \theta^2 = d\theta$, so there is a map of DG algebras from $W$ to $W'$ equipped with the total differential $d + \delta$ which sends $A$ to $\theta$ and $F$ to $d\theta$. As this map is an isomorphism, we obtain the desired bigraded structure on the Weil algebra.

In the sequel we will identify $W$ with $W'$ and use the bigraded notation for the Weil algebra. In this notation the Chern-Simons form of degree $2n+1$ for the connection $\theta$ and the universal trace $\natural : W \to W_\natural$ is

$$\mathrm{cs}_{2n+1} = \int_0^1 \natural\{\theta(td\theta + (t^2-t)\theta^2)^n\}dt/n! \qquad (22)$$

The transgression formula (3) it satisfies is

$$(d+\delta)\mathrm{cs}_{2n+1} = \natural\{d\theta^{n+1}\}/(n+1)! \qquad (23)$$

Writing the Chern-Simons form in homogeneous components for the bigrading of $W_\natural$ gives

$$\mathrm{cs}_{2n+1} = \sum_{p+k=n} \mathrm{cs}_{p,p+2k+1} \qquad (24)$$

$$\mathrm{cs}_{p,p+2k+1} = \frac{k!}{(p+2k+1)!} \natural\{\theta P_{pk}(d\theta, -\theta^2)\} \qquad (25)$$

where $P_{pk}(X, Y)$ denotes the sum of all the monomials in the noncommuting variables $X, Y$ which are of degree $p$ in $X$ and of degree $k$

in $Y$. Taking homogeneous components of the transgression relations (23) for different $n$ yields

$$d\{cs_{p,p+2k+1}\} + \delta\{cs_{p+1,p+2k}\} = 0 \text{ for } k \geq 1 \qquad (26)$$

$$d\{cs_{p,p+1}\} = \natural\{d\theta^{p+1}\}/(p+1)! \qquad (27)$$

## 2.5 Cohomology

Because of the importance of $W_\natural$ in constructing cyclic cohomology classes it is natural to study the various kinds of cohomology groups which can be associated to this double complex.

We first consider the $d$ cohomology. As far as the differential $d$ is concerned, $W$ is the tensor algebra of the complex with basis $\theta$, $d\theta$, and $W_\natural$ is the direct sum of the tensor powers of this complex divided by cyclic group actions. Since this complex is acyclic, it follows that the $d$ cohomology is zero.

Thus the double complex $W_\natural$ has exact rows, and so by the spectral sequence of a double complex the cohomology of the total complex with the differential $d + \delta$ is zero.

In order to find the $\delta$ cohomology, we first relate $W$ to the bar construction of a certain DG algebra $L$: $L_0 = \mathbf{C}1$, where 1 is the identity, $L_1 = \mathbf{C}s$, $L_n = 0$ in other degrees, $d(s) = -1$, and $s^2 = 0$. The bar construction $\bar{B}(L)$ (without counit) is a bigraded differential coalgebra which is the tensor coalgebra generated by $L$ located on the line of degree $q = 1$. The dual bigraded differential algebra $\bar{B}(L)^*$ is thus a tensor algebra with generators of degrees $(0,1)$, $(1,1)$. By checking the differentials one sees that $\bar{B}(L)^*$ and $W$ can be identified. Hence $W_\natural$ is the dual of $\bar{B}(L)^\natural$ the cyclic complex of $L$.

The differentials $d$ and $\delta$ in $\bar{B}(L)^\natural$ come from the differential and multiplication in $L$ respectively. Hence, the $\delta$ homology of $\bar{B}(L)^\natural$ is the same as the cyclic homology of the graded algebra $L$ with differential set to zero. This algebra is the result of adjoining an identity to the algebra $\mathbf{C}s$ with zero multiplication, and one knows [3, 4.3] in this case that the cyclic homology is the sum of the cyclic homologies of $\mathbf{C}$ and $\mathbf{C}s$. Hence the $\delta$ homology of $\bar{B}(L)^\natural$ is one dimensional in degrees $(0, 2k-1)$ and $(k,k)$ for $k \geq 1$, and zero elsewhere. By duality the $\delta$ cohomology of $W_\natural$ is located in the same places, and since $W$ is one dimensional in these degrees, we see that the classes of the elements $\natural(\theta^{2k-1})$ and $\natural(d\theta)^k$ form a basis for the $\delta$ cohomology.

The DG algebra $L$ is a very special case of a DG algebra of the form $\{I \to A\}$, where $I$ is an ideal in an algebra $A$. In [5] the bar construction and cyclic complex of such a DG algebra are studied, and the reader is referred there for further details.

## 9. Chern–Simons Forms and Cyclic Cohomology

We next consider the quotient algebra $W/I^m$, where $I$ is the ideal in $W$ generated by the curvature $d\theta$, and we compute the total cohomology of $(W/I^m)_\natural$. We know from the transgression formula that the Chern-Simons forms $cs_{2n-1}$ for $n \geq m$ are $d+\delta$ cocycles in this double complex. Now $W/I^m$ is obtained by killing the columns in $W$ with $p \geq m$, and similarly for its commutator quotient space. To compute the total cohomology of $(W/I^m)_\natural$, we use the spectral sequence starting with the $\delta$ cohomology. This has for basis the classes of $\natural(\theta^{2k-1})$ for $k \geq 1$ and $\natural(d\theta)^k$ for $1 \leq k < m$. The first $m-1$ of the former transgress to kill the latter classes, since this is true in $W_\natural$. Hence the classes of $\natural(\theta^{2k-1})$ for $k \geq m$ form a basis for $E^\infty$, and we find that the Chern-Simons classes $cs_{2n-1}$ for $n \geq m$ form a basis for the total cohomology of $(W/I^m)_\natural$.

On the other hand, let us consider the other spectral sequence associated to this double complex which starts with the $d$ cohomology. Using the fact that $W_\natural$ has exact rows, we see this spectral sequence collapses yielding

$$H^{p+q}((W/I^{p+1})_\natural) = H^q(W_\natural^{p,\cdot}/dW_\natural^{p-1,\cdot}) \qquad (28)$$

The cohomology on the right is the so-called $\delta$ mod $d$ cohomology of $W_\natural$ in degree $(p,q)$. This sort of cohomology (for a different bigraded algebra) appears in the study of anomalies [2]. By (26) the Chern-Simons components $cs_{p,p+2k+1}$ for $p \geq 0$, $k \geq 0$ are $\delta$ mod $d$ cocycles. From (28) and our calculation of the left side, the classes of these cocycles form a basis for the $\delta$ mod $d$ cohomology.

We summarize our calculations as follows.

**Theorem 1** *1) The $d$ cohomology and total cohomology of $\bar{W}_\natural$ are zero.*
*2) The $\delta$ cohomology is zero in all degrees except $(0, 2k-1)$ and $(k, k)$ for $k \geq 1$, where it is one dimensional with generator the classes of $\natural(\theta^{2k-1})$ and $\natural(d\theta^k)$ respectively.*
*3) The total cohomology of $(W/I^m)_\natural$, which is the quotient of $W_\natural$ obtained by killing the columns with $p \geq m$, has for basis the classes of the Chern-Simons forms $cs_{2n-1}$ for $n \geq m$.*
*4) The $\delta$ mod $d$ cohomology is zero in all degrees except $(p, p+2k+1)$ for $p \geq 0$, $k \geq 0$, where it is one dimensional with generator the class of $cs_{p,p+2k+1}$.*

The preceding discussion is motivated by the treatment of BRS algebras in [2]. One expects that there are very interesting relations between cyclic and BRS cohomology, but little is known at present.

## 3. Cyclic cohomology

In this section we first give an account of the theory of algebra cochains [4], which is then combined with Chern-Simons forms to construct cyclic cocycles.

All complexes we consider are **Z**-graded, and we use either upper or lower indexing related by the rule $K_{-n} = K^n$. The differential is of degree $-1$ in the lower indexing, degree $+1$ in the upper.

Given complexes $K$, $L$, let $\mathrm{Hom}(K,L)^{pq} = \mathrm{Hom}(K_q, L^p)$; this is a double complex with differentials $df = d_L \cdot f$, $\delta f = (-1)^{|f|+1} f \cdot d_K$, where $|f| = p + q$ is the total degree. The total complex with the differential $d_{tot} = d + \delta$ is the complex of linear maps from $K$ to $L$, provided $K_n = L^n = 0$ for $n \ll 0$, which is true in the cases we consider.

### 3.1 Bar construction and cochains

Let $A$ be an algebra (without identity), and let $\bar{B}(A)$ be the complex with $\bar{B}(A)_n = A^{\otimes n}$ for $n \geq 1$, with zero in other degrees, and with differential

$$b'(a_1,\ldots,a_n) = \sum_{i=1}^{n-1} (-1)^{i-1}(a_1,\ldots,a_i a_{i+1},\ldots,a_n) \qquad (29)$$

If $\Delta : A \to A \otimes A$ is the coproduct given by

$$\Delta(a_1,\ldots,a_n) = \sum_{i=1}^{n-1} (a_1,\ldots,a_i) \otimes (a_{i+1},\ldots,a_n) \qquad (30)$$

one can verify that $\bar{B}(A)$ is a DG coalgebra. It is the bar construction (without counit) of $A$.

If $L$ a DG algebra, then $R = \mathrm{Hom}(\bar{B}(A), L)$ is naturally a bigraded differential algebra. An element of degree $(p', p)$ is a multilinear function $f(a_1,\ldots,a_p)$ on $A$ with values in $L^{p'}$; we say $f$ has $A$-degree $p$ and $L$-degree $p'$. The two differentials are

$$\delta f = (-1)^{|f|+1} f \cdot b' \qquad df = d_L \cdot f \qquad (31)$$

The product is $fg = m(f \otimes g)\Delta$, where $m$ is the multiplication in $L$ and where $f \otimes g$ is the tensor product of maps of complexes involving the usual signs. Hence

$$(fg)(a_1,\ldots,a_{p+q}) = (-1)^{p|g|} f(a_1,\ldots,a_p) g(a_{p+1},\ldots,a_{p+q}) \qquad (32)$$

if $g$ has $A$-degree $q$.

For example, an element of $R$ of degree $(0,1)$ is a linear map $\rho : A \to L^0$. One has

$$(\delta \rho + \rho^2)(a_1, a_2) = \rho(a_1 a_2) - \rho(a_1)\rho(a_2) \qquad (33)$$

## 9. Chern–Simons Forms and Cyclic Cohomology

showing that $\delta\rho + \rho^2$ vanishes if and only if $\rho$ is an algebra homomorphism.

### 3.2 Cyclic formalism

Let $B^c(A)$ denote the complex with $B^c(A)_n = A^{\otimes n}$ for $n \geq 1$, with zero in other degrees, and with differential

$$b(a_1,\ldots,a_n) = b'(a_1,\ldots,a_n) + (-1)^{n-1}(a_n a_1, a_2,\ldots,a_{n-1}) \qquad (34)$$

It is essentially the cyclic bar construction of $A$, or equivalently, the standard complex giving the Hochschild homology of $A$ with coefficients in $A$ considered as a bimodule. The only difference is a harmless shift in degrees which we make in order to fit the grading on the bar construction.

Let $\lambda$, $N$ be the operators on $A^{\otimes n}$ given by

$$\lambda(a_1,\ldots,a_n) = (-1)^{n-1}(a_n, a_1,\ldots,a_{n-1}), \qquad N = \sum_0^{n-1} \lambda^i \qquad (35)$$

It is a basic fact of cyclic theory that these maps for different $n$ fit together to give maps of complexes

$$\bar{B}(A) \xrightarrow{\lambda-1} B^c(A) \xrightarrow{N} \bar{B}(A) \qquad (36)$$

In other words we have the identities

$$b(\lambda-1) = (\lambda-1)b' \qquad b'N = Nb \qquad (37)$$

Tsygan's proof of these identities uses the following formulas for $b'$, $b$ on $A^{\otimes n}$

$$b' = \sum_{j=1}^{n} \lambda^{j-1} c \lambda^{-j} \qquad b = \sum_{j=1}^{n+1} \lambda^{j-1} c \lambda^{-j} \qquad (38)$$

where $c(a_1,\ldots,a_n) = (-1)^{n-1}(a_n a_1, a_2,\ldots,a_{n-1})$ is the crossover term in the formula for $b$. It is straightforward to verify (38) and to derive (37) from it.

Because $\mathrm{Ker}(\lambda-1) = \mathrm{Im}(N)$ and $\mathrm{Ker}(N) = \mathrm{Im}(\lambda-1)$, we can build an exact sequence of complexes

$$\xrightarrow{N} \bar{B}(A) \xrightarrow{\lambda-1} B^c(A) \xrightarrow{N} \bar{B}(A) \xrightarrow{\lambda-1} \qquad (39)$$

which can be viewed as a double complex which is periodic of period two in the horizontal direction. It is obtained by splicing together copies of the exact sequence of complexes

$$0 \longrightarrow \bar{B}(A)^\natural \xrightarrow{\natural} \bar{B}(A) \xrightarrow{\lambda-1} B^c(A) \xrightarrow{N} \bar{B}(A)^\natural \longrightarrow 0 \qquad (40)$$

where $\bar{B}(A)^{\natural}$ is the subcomplex $\text{Im}(N) = \text{Ker}(\lambda - 1)$ and $\natural$ is the inclusion.

By exactness $\bar{B}(A)^{\natural}$ is isomorphic to the cokernel of $\lambda - 1$. This cokernel is by definition the complex giving the cyclic homology of $A$ except for the degree shift we have mentioned, so we have

$$H_n(\bar{B}(A)^{\natural}) = HC_{n-1}(A) \tag{41}$$

In [4] it is shown how (39), (40) can be interpreted as coalgebra analogues of the sequences (11), (12) for a DG algebra. In particular $\bar{B}(A)^{\natural}$ is the cocommutator subspace of $\bar{B}(A)$, that is, the coalgebra analogue of $R/[R, R]$.

*3.3  A map of periodic complexes*

We are interested in cyclic cohomology, and so we want to consider cochains corresponding to the complexes $\bar{B} = \bar{B}(A)$, $B^c = B^c(A)$, and $\bar{B}^{\natural} = \bar{B}(A)^{\natural}$. Let $V$ be a complex with $V^n = 0$ for $n \ll 0$, and let us write $K^*$ for $\text{Hom}(K, V)$. Elements of $\bar{B}^*$, $B^{c,*}$, and $\bar{B}^{\natural,*}$ will be referred to as bar, Hochschild, and cyclic cochains respectively. The cohomology of $\bar{B}^{\natural,*}$ is the cyclic cohomology with coefficients $V$.

Applying $*$ to (39), (40) we get exact sequences of complexes of cochains

$$\xrightarrow{(\lambda-1)^t} \bar{B}^* \xrightarrow{N^t} B^{c,*} \xrightarrow{(\lambda-1)^t} \bar{B}^* \xrightarrow{N^t} \tag{42}$$

$$0 \longrightarrow \bar{B}^{\natural,*} \xrightarrow{N^t} B^{c,*} \xrightarrow{(\lambda-1)^t} \bar{B}^* \xrightarrow{\natural^t} \bar{B}^{\natural,*} \longrightarrow 0 \tag{43}$$

where $t$ stands for transpose. Diagram chasing in (43) as in (13) yields the $S$-operation on the cyclic cohomology which raises degree by two.

Let $\tau : L \to V$ be a trace on $L$. We are going to show that it determines a map from the periodic complex (11) of $R = \text{Hom}(\bar{B}, L)$ to the periodic complex (42) of cochains on $A$, hence also a map from (12) to (43). This will show then that the $S$-relations we established before imply $S$-relations in cyclic cohomology.

Applying $\tau$ to the values of cochains gives a map

$$\tau : R \to \bar{B}^* \qquad \tau(f) = \tau \cdot f \tag{44}$$

compatible with both differentials. It is not a trace on $R$ but satisfies

$$\tau(fg)\lambda^p = (-1)^{|f||g|}\tau(gf) \tag{45}$$

where $p$ is the $A$-degree of $f$. To see this, suppose $f, g$ of degrees

## 9. Chern–Simons Forms and Cyclic Cohomology

$(p', p)$, $(q', q)$. Then

$$\begin{aligned}
\tau(fg)\lambda^p(a_1,\ldots,a_{q+p}) &= (-1)^{pq}\tau(fg)(a_{q+1},\ldots,a_{q+p},a_1,\ldots,a_q) \\
&= (-1)^{pq}(-1)^{p(q+q')}\tau\{f(a_{q+1},\ldots,a_{q+p})g(a_1,\ldots,a_q)\} \\
&= (-1)^{pq'}(-1)^{p'q'}\tau\{g(a_1,\ldots,a_q)f(a_{q+1},\ldots,a_{q+p})\} \\
&= (-1)^{(p+p')q'}(-1)^{q(p+p')}\tau(gf)(a_1,\ldots,a_{q+p})
\end{aligned}$$

Composing with the inclusion $\natural : \bar{B}^\natural \to \bar{B}$, which gives the kernel of $\lambda - 1$, we get a map

$$\tau^\natural : R \to \bar{B}^{\natural,*} \qquad \tau^\natural(f) = \tau f \natural \qquad (46)$$

which is a trace on $R$ with values in cyclic cochains.

We next construct a map from $\Omega^1_{R,\natural}$ to Hochschild cochains.
Given $f, g \in R$ as above, let

$$\psi(f,g) = \tau(fg)\sum_0^{p-1}\lambda^i \qquad (47)$$

This is a map $A^{\otimes p+q} \to V^{p'+q'}$ which will be viewed as an element of the complex $B^{c,*}$ of Hochschild cochains. We extend the definition in the obvious way

$$\psi(f,1) = \tau(f)\sum_0^{p-1}\lambda^i = \tau f N \qquad (48)$$

so that $\psi(f,g)$ is defined for $g \in \tilde{R}$.

We will show the pairing $\psi$ has the following properties:

$$\begin{aligned}
\delta\{\psi(f,g)\} &= \psi(\delta f, g) + (-1)^{|f|}\psi(f, \delta g) & (49) \\
d\psi(f,g) &= \psi(df, g) + (-1)^{|f|}\psi(f, dg) & (50) \\
\psi(fg, h) &= \psi(f, gh) + (-1)^{|f||gh|}\psi(g, hf) & (51)
\end{aligned}$$

Equation (50) results from $d\tau = \tau d$, and (51) follows from

$$\begin{aligned}
\tau(fgh)\sum_0^{p+q-1}\lambda^i &= \tau(fgh)\sum_0^{p-1}\lambda^i + \tau(fgh)\lambda^p\sum_0^{q-1}\lambda^i \\
&= \psi(f,gh) + (-1)^{|f||gh|}\tau(ghf)\sum_0^{q-1}\lambda^i
\end{aligned}$$

using (45). This also holds when $h = 1$.

We next prove (49). When $g = 1$, it results from the identity $b'N = Nb$. When $g \in R$, the proof is a more complicated version of the proof of this identity using the same tools (38). By the definition of $\delta, \psi$ we have

$$\delta\{\psi(f,g)\} = (-1)^{|f|+|g|+1}\psi(f,g)b$$
$$= (-1)^{|f|+|g|+1}\tau(fg)\sum_{k=0}^{p-1}\lambda^k\sum_{l=1}^{p+q+1}\lambda^{l-1}c\lambda^{-l}$$
$$\psi(\delta f,g) = (-1)^{|f|+1}\tau\{(fb')g\}\sum_{i=0}^{p}\lambda^i$$
$$(-1)^{|f|}\psi(f,\delta g) = (-1)^{|f|+|g|+1}\tau\{f(gb')\}\sum_{i=0}^{p-1}\lambda^i$$

and using (32), (38) we have

$$\tau\{(fb')g\} = (-1)^{|g|}\tau(fg)\sum_{j=0}^{p}\lambda^{j-1}c\lambda^{-j}$$

$$\tau\{f(gb')\} = \tau(fg)\sum_{j=p+1}^{p+q}\lambda^{j-1}c\lambda^{-j}$$

so it suffices to prove the identity

$$\sum_{k=0}^{p-1}\sum_{l=1}^{p+q+1}\lambda^{k+l-1}c\lambda^{-l} = \left\{\sum_{j=1}^{p}\sum_{i=0}^{p}+\sum_{j=p+1}^{p+q}\sum_{i=0}^{p-1}\right\}\lambda^{j-1}c\lambda^{-j+i} \quad (52)$$

Note that on each side we have $p(p+q+1)$ terms. We show that both sides consist of the same terms indexed differently. Recall that the $\lambda$ on the left of $c$ belongs to a cyclic group of order $(p+q)$, and that the order is one more for the $\lambda$ on the right. Given a pair $k, l$ on the left, the corresponding pair $i, j$ is given by

$$\begin{array}{llll} i = k, & j = k+l & \text{if} & k+l \leq p+q \\ i = k+1, & j = k+l-(p+q) & \text{if} & k+l > p+q \end{array}$$

The inverse correspondence is

$$\begin{array}{llll} k = i, & l = j-i & \text{if} & i < j \\ k = i-1, & l = (p+q+1)-(i-j) & \text{if} & i \geq j \end{array}$$

This completes the proof of (49).

By the property of $\Omega^1_{R,\natural}$ with respect to the cocycle condition (51) described in §1.2, we obtain a map

$$\tau^\natural : \Omega^1_{R,\natural} \longrightarrow B^{c,*} \qquad \tau^\natural(\partial fg) = \psi(f,g) \quad (53)$$

## 9. Chern–Simons Forms and Cyclic Cohomology

which is compatible with the differentials $\delta, d$ by (49), (50).

We next verify that (44) and (52) constitute a map from the periodic complex (11) of $R$ to the periodic complex of bar and Hochschild cochains (42), that is, we have a commutative diagram

$$
\begin{array}{ccccc}
R & \xrightarrow{\bar{\partial}} & \Omega^1_{R,\natural} & \xrightarrow{\beta} & R \\
\downarrow{\tau} & & \downarrow{\tau^\natural} & & \downarrow{\tau} \\
\bar{B}^* & \xrightarrow{N^t} & B^{c,*} & \xrightarrow{(\lambda-1)^t} & \bar{B}^*
\end{array}
\qquad (54)
$$

The commutativity of the first square follows from

$$\tau^\natural(\partial f) = \psi(f,1) = \tau f N = N^t(\tau f)$$

As for the second square we have using (45)

$$(\lambda-1)^t \tau^\natural(\partial f g) = \tau(fg) \sum_0^{p-1} \lambda^i(\lambda-1) = \tau(fg)(\lambda^p - 1) =$$

$$(-1)^{|f||g|}\tau(gf) - \tau(fg) = \tau(-[f,g]) = \tau\{\beta(\partial f g)\}$$

### 3.4 Applications to cyclic cohomology

We consider the following situation: $A$ is an algebra, $L$ is a DG algebra, and $\tau : L \to V$ is a trace on $L$. On the bigraded differential algebra $R = \mathrm{Hom}(\bar{B}(A), L)$ we have the trace (46) with values in the complex $\mathrm{Hom}(\bar{B}(A)^\natural, V)$ of cyclic cochains. Thus we are in a position to produce cyclic cohomology classes on $A$ by working in $R$.

Suppose given an algebra homomorphism $A \to L^0$. This may be interpreted as an element $\theta$ of $R$ of degree $(0,1)$ satisfying $\delta\theta + \theta^2 = 0$. Using $\theta$ as 'connection' we construct the Chern-Simons form (4), (22)

$$\mathrm{cs}_{2n-1} \in \mathrm{Hom}(\bar{B}(A)^\natural, V) \qquad (55)$$

This is a cyclic cochain satisfying

$$(d+\delta)\mathrm{cs}_{2n-1} = \tau\{d\theta^n/n!\}_\natural \qquad (56)$$

We assume $V^p = 0$ for $p \geq m$. When $n \geq m$, the right side is then zero, and $\mathrm{cs}_{2n-1}$ is a cyclic cocycle.

In the previous section we produced a map from the periodic complex (11) of $R$ to the periodic complex (42) of bar and Hochschild cochains on $A$. Consequently the $S$-relations (20) established in §1.3 using the former carry over to the latter to show that the cyclic cohomology classes $[\mathrm{cs}_{2n-1}]$ for $n \geq m$ satisfy

$$S[\mathrm{cs}_{2n-1}] = [\mathrm{cs}_{2n+1}] \qquad (57)$$

for the $S$ operation on cyclic cohomology.

We now wish to describe the cyclic cochains given by the Chern-Simons forms concretely, and for this purpose we need some notation.

Let $CC(A)$ denote the standard cyclic complex with $CC_n(A) = A^{\otimes n+1}_\lambda$, the quotient for the cyclic action given by $\lambda$, and with differential $b$. The map $N$ identifies $CC(A)$ and $\bar{B}(A)^\natural$ except for the difference in conventions about how the degrees are defined. Given $f \in R = \mathrm{Hom}(\bar{B}(A), L)$ of degree $(p,q)$, we can therefore identify $f\natural : \bar{B}(A)^\natural_q \to L^p$ with the cyclic cochain of degree $q-1$

$$fN(a_1,\ldots,a_q) = \sum_{i=0}^{q-1}(-1)^{i(q-1)} f(a_{q-i+1},\ldots,a_q,a_1,\ldots,a_{q-i}) \qquad (58)$$

We consider the universal situation where $L$ is the DG algebra $\Omega = \Omega_A$ of noncommutative differential forms over $A$ [1]. It is the universal DG algebra generated by $A$ in degree zero and is such that

$$\tilde{A} \otimes A^{\otimes n} \cong \Omega^n_A \qquad \tilde{a}_0 \otimes a_1 \otimes \ldots \otimes a_n \leftrightarrow \tilde{a}_0 da_1 \ldots da_n \qquad (59)$$

We take $\tau$ to be the canonical surjection of $\Omega_A$ onto $\Omega_{A,\natural}$, and take the homomorphism $A \to \Omega^0_A$ to be the identity.

In this notation the Chern-Simons form component (25) is

$$\mathrm{cs}_{p,p+2k+1} = \frac{k!}{(p+2k+1)!} \natural\{\theta P_{pk}(d\theta, -\theta^2)\} N \qquad (60)$$
$$\in \mathrm{Hom}(CC_{p+2k}(A), \Omega^p_{A,\natural})$$

which is a cyclic cochain of degree $p+2k$ evaluated as follows. When $\theta P_{pk}(d\theta, -\theta^2)$ is applied to $(a_0,\ldots,a_{p+2k})$ there are no signs, because $d\theta$, $\theta^2$ are of even degree and $-\theta^2(a_1,a_2) = a_1 a_2$. Thus in each monomial occurring in $P_{pk}$ one substitutes either $da_i$ or $a_i a_{i+1}$ keeping the $a_i$ in order, e.g.

$$(\theta d\theta(-\theta^2) d\theta)(a_0,\ldots,a_4) = a_0 da_1 a_2 a_3 da_4 \qquad (61)$$

We now consider instead of the universal trace $\natural$ on $\Omega$ a trace $\Omega \to \mathbf{C}[p]$, where $\mathbf{C}[p]$ is the complex with $\mathbf{C}$ in degree $p$ and zero elsewhere. Such a trace is given by a linear functional $z$ on $\Omega^p$ vanishing on $d\Omega^{p-1} + [\Omega,\Omega]^p$; it is a closed trace of degree $p$ on $\Omega$ in Connes sense, which is sometimes denoted

$$\omega \mapsto \int_z \omega \qquad \omega \in \Omega^p \qquad (62)$$

to emphasize the analogy with integration over a closed current $z$. We construct the Chern-Simons form of degree $2n+1$ with this trace. This

## 9. Chern–Simons Forms and Cyclic Cohomology

gives zero for $n < p$, and for $n = p+k$, $k \geq 0$ we obtain the (ordinary C-valued) cyclic cochain of degree $p+2k$

$$\int_z \mathrm{cs}_{p,p+2k+1} = \frac{k!}{(p+2k+1)!} \int_z \theta P_{pk}(d\theta, -\theta^2) N \tag{63}$$

We have seen that these are cyclic cocycles, and that their cyclic cohomology classes are related by the $S$-operation. We now want to understand this result better and to relate it to Connes' development of cyclic cohomology via noncommutative differential forms [1].

In concrete terms the reason (63) is closed is the following. From the transgression relation (26) we have

$$d\,\mathrm{cs}_{p-1,p+2k+2} + \mathrm{cs}_{p,p+2k+1}\,b = 0 \tag{64}$$

for $k \geq 0$. Hence $b$ applied to (63) is in the image of $\int_z d$ which is zero.

Incidentally, if $z$ is a 'boundary' in the sense that

$$\int_z \omega = \int_y d\omega \tag{65}$$

for some linear functional $\int_y$ on $\Omega_{\natural}^{p+1}$, then for $k > 0$ we have

$$\int_z \mathrm{cs}_{p,p+2k+1} = \int_y d\mathrm{cs}_{p,p+2k+1} = \left(\int_y \mathrm{cs}_{p+1,p+2k}\right) b \tag{66}$$

showing that the cyclic cocycle is a cyclic coboundary. Thus for $k > 0$ the cyclic cohomology class of (63) depends only on the 'de Rham homology class' of $z$.

According to Connes' basic proposition [1], II, Proposition 1, which is the starting point of his theory of cyclic cohomology and noncommutative differential forms, the cochain

$$\frac{1}{p!}\int_z (\theta d\theta^p)(a_0,\ldots,a_p) = \frac{1}{p!}\int_z a_0 da_1 \ldots da_p \tag{67}$$

is a cyclic $p$-cocycle. Moreover, this formula gives an equivalence between cyclic $p$-cocycles and closed traces of degree $p$ on $\Omega_A$. Because this cochain is already cyclic we have

$$\frac{1}{(p+1)!}\int_z (\theta d\theta^p) N = \frac{1}{p!}\int_z \theta d\theta^p \tag{68}$$

which identifies (63) for $k = 0$ with Connes' basic cyclic cocycle (67).

Connes also gives the following explicit description of the $S$-operation on the level of cyclic cocycles. Let $\mathbf{C}e$, $e^2 = e$ denote the complex numbers considered as a nonunital algebra. Form $\Omega_{\mathbf{C}e}$ and let $\tau_{2k}$ be the closed trace of degree $2k$ on it given by $\tau_{2k}(ede^{2k}/k!) = 1$, $\tau_{2k}(de^{2k}) = 0$. By the universal property of $\Omega_A$ as a DG algebra, there is a unique map of DG algebras

$$\pi : \Omega_A \to \Omega_A \otimes \Omega_{C_e} \qquad (69)$$

such that $\pi(a) = a \otimes e$. Given a closed trace $\int_z$ of degree $p$ on $\Omega_A$, we compose the tensor product of $\int_z$ and $\tau_{2k}$ with $\pi$ to get a closed trace of degree $p + 2k$. Since closed traces on $\Omega_A$ are equivalent to cyclic cocycles, we get an operation raising degree by $2k$ on cyclic cocycles. This is the Connes description of $S^k$.

We now calculate what $S^k$ does to the cyclic cocycle (67). Abbreviating $\omega \otimes \eta$ to $\omega\eta$, we have $\pi(d\theta) = d(\theta e) = X + Y$, where $X = d\theta e$, $Y = -\theta de$. Also

$$(\int_z \tau_{2k})\pi(\theta d\theta^n) = \int_z \tau_{2k}\{\theta e(X+Y)^n e\} \qquad (70)$$

since $e = e^2$ and we can use the trace property to move one $e$ around to the right. Because $e\,de\,e = 0$, $de^2 e = ede^2$, we have $eY^{2i+1}e = 0$. So when we expand $e(X+Y)^n e$, only monomials in $X$ and $Y^2 = -\theta^2 de^2$ contribute, and we find

$$e(X+Y)^n e = \sum_{p+2k=n} eP_{pk}(X,Y^2)e = \sum_{p+2k=n} P_{pk}(d\theta, -\theta^2)\,ede^{2k} \qquad (71)$$

Thus $S^k$ applied to (67) is

$$\frac{1}{(p+2k)!}(\int_z \tau_{2k})\pi(\theta d\theta^{p+2k}) = \frac{k!}{(p+2k)!}\int_z \{\theta P_{pk}(d\theta, -\theta^2)\} \qquad (72)$$

This calculation shows that the cochain on the right is already cyclic, and consequently that the cyclic cocycle (63) obtained via Chern-Simons forms agrees with the one (72) obtained via Connes method.

### References

[1] A. Connes, Non-commutative differential geometry, *Publ. Math. I. H. E. S.* **62** (1985), 41-144.

[2] M. Dubois-Violette, M. Talon, and C. M. Viallet, B.R.S. Algebras. Analysis of the consistency equations in gauge theory, *Commun. Math. Phys.* **102** (1985), 105-122.

[3] J.-L. Loday and D. Quillen, Cyclic homology and the Lie algebra of matrices, *Comment. Math. Helv.* **59**(1984), 565-591.

[4] D. Quillen, Algebra cochains and cyclic cohomology, *Publ. Math. I. H. E. S.* **68** (1989), 139-174.

[5] D. Quillen, Cyclic cohomology and algebra extensions, *K-Theory*, to appear.

# 10

## A Yang-Mills Structure for String Field Theory

*Tsou Sheung Tsun*

String theorists believe that one way to achieve a fully quantized theory of strings is through string field theory. The other way is to study conformal field theory on Riemann surfaces of different genera, which is the subject of many of the talks at this Conference. In a way, string field theory is the more conservative approach, since it aims just to replace the spacetime points of conventional quantum field theory by strings, which are extended objects. However, from this point of view string theory has one rather unsatisfactory aspect, in the sense that although it has been very well developed and minutely studied, we are still rather unclear about its basic structure. We can contrast this to both general relativity, which is based on the geometry of spacetime, and to gauge theory, which is about the structure of various natural bundles over spacetime. And yet string theory is supposed to embody both these two essentially geometric theories. To paraphrase Witten [1], in string theory we seem to have to work backwards to get at the still unknown basic structure. In this talk I would like to report on some joint work [2,3] with Chan Hong-Mo in an attempt to gain some understanding in that general direction. It seems that one could in some sense consider string field theory as a generalized Yang-Mills theory.

Our starting point is Witten's paper on string theory and non-commutative geometry [4]. To introduce the notations let me summarize it very briefly. Witten postulates a non-commutative differential graded algebra $\mathcal{A}$, of which the 1-forms are the string fields. He denotes the wedge-product by $\star$, and $\mathcal{A}$ is generally referred to as Witten's star algebra. Commutativity is taken in the graded sense, as in Connes [5], so that we have in general

$$\omega_p \star \omega_q \neq (-1)^{pq} \omega_q \star \omega_p,$$

for $\omega_p$ a $p$-form and $\omega_q$ a $q$-form. The exterior derivative is the BRST operator $Q$. There is a linear functional on $\mathcal{A}$, denoted by $\int$, which is in fact a combination of integration and trace. With these ingredients, string field theory is built up from an action which is the integral of

a Chern-Simons form:
$$I = \int A \star QA + \frac{2}{3} A \star A \star A. \tag{1}$$

This action is invariant under the transformation:
$$\delta A = Q\varepsilon + A \star \varepsilon - \varepsilon \star A, \tag{2}$$

where $\varepsilon$ is an arbitrary 0-form. To make contact with the Veneziano amplitudes, the integrals are actually evaluated as path integrals on the string worldsheet.

As pointed out by Witten all along, the elements of the $\star$-algebra behave much as infinite-matrix-valued differential forms. We wish to make this a little more explicit, and to show how a Yang-Mills structure can emerge.

Now in the original 1954 paper[6] by Yang and Mills, gauge invariance was described as the requirement that all interactions be invariant under independent gauge transformations at all spacetime points. In other words, the relative non-abelian phase of the wave function before and after a spacetime translation has no physical meaning. Suppose that, instead of the group of spacetime translations, we consider some other group $B$ which acts on the wave functions. It may then happen that the relative non-abelian phase of the wave function before and after the action of an element of $B$ has no physical meaning. In this case, one may try to construct a Yang-Mills theory based on $B$ rather than the usual spacetime translations [7]. This is not an entirely idle speculation. In string theory, the string wave functions are actually functionals, and they are acted on by the group of worldsheet reparametrizations, variously known as the Virasoro group or Diff $S^1$. Further, the physics is invariant under this group. The passage from the finite-dimensional abelian translation group to the infinite-dimensional non-abelian Virasoro group entails significant modifications which I shall spell out later.

The other important ingredient in a Yang-Mills theory is the gauge group. In searching for what this gauge group ought to be, let us first try to draw some simple-minded analogies with a known theory. Originally the non-abelian gauge group is the matrix group $SU(2)$ representing symmetry in some internal space unconnected to spacetime. Call this isospin for convenience. Immediately we see in (2) an analogue to this usual gauge invariance:
$$\delta A_\mu = \partial_\mu \varepsilon - [A_\mu, \varepsilon], \tag{3}$$

where $\varepsilon(x)$ here is a hermitian matrix. Now a string has an extension $X(\sigma)$, $\sigma = 0 \to \pi$, which naturally gives rise to *intrinsic* internal

## 10. A Yang-Mills Structure for String Field Theory

degrees of freedom directly connected to spacetime.[1] If we are to build a gauge theory then these internal degrees of freedom ought to be gauged. By analogy with isospin above, we look for a group of unitary transformations. Now the string fields are 1-forms and so are the analogues of the gauge potential, which is in the adjoint representation of the unitary gauge group. To pursue the analogy we must then interpret the string field functional $A[X(\sigma)]$, $\sigma = 0 \to \pi$ as a "hermitian matrix". The most simple-minded way to do this is to split the string into two half-strings, called "commas" in [2]. The left half-string, with $\sigma = 0 \to \pi/2$, represents column indices and the right half-string, with $\sigma = \pi/2 \to \pi$, represents row indices. The use of the continuum indices here is, I am afraid, not all that well-defined, but there is a better, although less intuitive, way of going about it which I shall describe shortly. So in this picture a functional $\Psi[X_L(\sigma)]$ of a half-string ($\sigma = 0 \to \pi/2$) is then a vector indexed by the points $X_L(\sigma)$ on the half-string, and a full-string functional $A[X(\sigma)]$, $\sigma = 0 \to \pi$, can then be regarded as a matrix with column indices $X_L(\sigma)$, $\sigma = 0 \to \pi/2$ and row indices $X_R(\pi - \sigma)$, $\sigma = \pi/2 \to \pi$. In fact the half-string functional is something like a QCD quark in the fundamental representation, and the 1-form full-string functional is like the gluon (gauge) field in the adjoint representation. The hermicity condition can now be written in half-string matrix form as:

$$A^\dagger[X_1, X_2] = A[X_1, X_2], \tag{4}$$

where by definition $A^\dagger[X_1, X_2] = \overline{A[X_2, X_1]}$. The hermicity condition (4), when translated into full-string language, is identical to Witten's reality condition [8]

$$\overline{A[X]} = A[\overline{X}], \tag{5}$$

where by definition $\overline{X}(\sigma) = X(\pi - \sigma)$. So we tentatively conclude that this unitary group of infinite matrices is the gauge group for our string theory. We note that of course we are as yet far from making any formal definitions.

The advantages of dealing with matrices are many. For instance, one can write down products and invariants like traces, at least formally. In the functional formalism, we can write them as follows:

$$\begin{aligned}(M \cdot N)[X_1, X_3] &= \int \delta X_2 M[X_1, X_2] N[X_2, X_3],\\ Tr M &= \int \delta X M[X, X]\end{aligned} \tag{6}$$

where integrals replace discrete summations formally.

---

[1] Any *extrinsic* isospin-like internal symmetry that the string may possess can be treated in the usual way using the Chan-Paton trace factors. At this stage we simplify the formalism by ignoring it.

However, to make better sense of matrix formulae such as (4) to (6), it is expedient to turn to an operator formalism. From our foregoing considerations we want to establish such a formalism based on half-strings; but first let us recall the full-string construction. The full-string obeys open boundary conditions:

$$\frac{\partial}{\partial \sigma} X^\mu(\tau, \sigma) = 0, \quad \text{at} \quad \sigma = 0, \pi, \tag{7}$$

where $\sigma$ represents the "position" and $\tau$ the "time" on the string worldsheet. Quantization leads to oscillator modes $\alpha_n$ with commutation relations:

$$[\alpha_n^\mu, \alpha_m^\nu] = -n\delta(n+m)\eta^{\mu\nu}, \tag{8}$$

where $\eta^{\mu\nu}$ is the Minkowski metric of spacetime. The Fock space is then built up by letting the $\alpha_{-n}$, $n > 0$, act on the full-string vacuum in the usual way. In contrast, the half-string obeys a different set of boundary conditions[2]:

$$\frac{\partial}{\partial \sigma} X^\mu(0, \tau) = 0, \quad \frac{\partial}{\partial \tau} X^\mu(\frac{\pi}{2}, \tau) = 0, \tag{9}$$

leading to oscillator modes $\beta_k$, $k$ odd only, with commutation relations:

$$[\beta_k^\mu, \beta_l^\nu] = -k\delta(l+k)\eta^{\mu\nu}. \tag{10}$$

We can also build up a Fock space, tensor product of the $\beta_{-k}$, $k > 0$, acting on half-string vacua (left and right), together with the free motion of the mid-point. The transformation of the $\alpha$-Fock space to the product $\beta$-Fock space, and *vice versa*, has been worked out explicitly, but is quite involved [9]. This is why, in spite of the fact that quantities are much better defined in an operator formalism, it is often clearer to present these ideas in the more intuitive functional formalism. As usual, one can go from one formalism to the other via Fourier transform.

Now we are in a position to have a closer look at the "base group" $B$ which replaces the group of spacetime translations. The Virasoro generators can be represented as:

$$L_n^{(\alpha)} = \frac{1}{2} \sum_m : \alpha_m \alpha_{n-m} :, \tag{11}$$

where the normal ordering is with respect to the $\alpha$ oscillator modes. Alternatively we can use the Fourier transform:

$$L_{\pm\sigma}^{(\alpha)} = \frac{1}{2} : \left\{ -i\pi \frac{\delta}{\delta X(\sigma)} \pm X'(\sigma) \right\}^2 :, \quad \sigma = 0 \to \pi, \tag{12}$$

---
[2]In order to present a clear picture without too many formulae we here gloss over subtleties involving the mid-point. The details can be found in [3].

## 10. A Yang-Mills Structure for String Field Theory

with the same normal ordering. In terms of the half-string, however, we should consider instead

$$L^{(\beta)}_{\pm\sigma} = \frac{1}{2} : \left\{-i\pi \frac{\delta}{\delta X(\sigma)} \pm X'(\sigma)\right\}^2 : \quad \sigma = 0 \to \pi/2, \quad (13)$$

where the normal ordering is with respect to the $\beta$ oscillator modes, with corresponding Fourier components $L_n^{(\beta)}$. It turns out [3] that it is the $L_\sigma^{(\beta)}$ or $L_n^{(\beta)}$ which generate the "base group" $B$ on which we can construct a Yang-Mills theory. (13) defines the action of $L_\sigma^{(\beta)}$ on half-string functionals $\Psi[X(\sigma)]$ inducing in the usual way an action on half-string matrices, which are full-string functionals. It can be shown [3] that this induced action coincides, apart from subtleties at the mid-point, with the action of the operators:

$$K_\sigma = L_\sigma^{(\alpha)} - L_{\pi-\sigma}^{(\alpha)}, \quad \sigma = 0 \to \pi/2; \quad (14)$$

or their Fourier components:

$$K_n = L_n^{(\alpha)} - (-1)^n L_{-n}^{(\alpha)}. \quad (15)$$

These particular combinations of the Virasoro operators were first pointed out by Witten [8] as constituting the invariance of the interacting string, this being a subgroup of the Virasoro group which constitutes the invariance of the free string[3].

The $L_\sigma$ or equivalently the $L_n$ (dropping the superscript $\beta$ whenever there is no cause for confusion) then take the place of the usual spacetime translation operators $\partial_\mu$. Dual to the $\partial_\mu$ are the 1-forms $dx^\mu$. Similarly one can construct the 1-forms dual to the $L_\sigma$, denoted by $\eta^\sigma$. They are the half-string analogues of the usual BRST ghosts. We have now all the ingredients needed to construct a Yang-Mills theory. We list the two sets of ingredients in the following table:

---

[3] See [3] for another interpretation of the $K_\sigma$.

|  | Yang-Mills Theory | String Theory |
|---|---|---|
| translation | $\partial_\mu$ | $L_\sigma$ |
| 1-forms | $dx^\mu$ | $\eta^\sigma$ |
| potential | $A = A_\mu dx^\mu$, $A_\mu^{ij}$ hermitian | $A = \int_{-\pi/2}^{\pi/2} A_\sigma \eta^\sigma d\sigma$, $A_\sigma[X_1, X_2]$ hermitian |
| gauge transformations (infinitesimal) | $\varepsilon^{ij}(x)$ | $\varepsilon[X_1, X_2]$ |
| change in wave function (infinitesimal) | $\delta\psi_i = \sum_i \varepsilon^{ij}(x)\psi_j(x)$ | $\delta\Psi[X_1] = \int \delta X_2 \varepsilon[X_1, X_2]\Psi[X_2]$ |
| change in potential (infinitesimal) | $\delta A_\mu = [(\partial_\mu - A_\mu), \varepsilon]$ | $\delta A_\sigma = [(L_\sigma - A_\sigma), \varepsilon]$ |
| exterior derivative | $d = \partial_\mu dx^\mu$ | $Q = \int_{-\pi/2}^{\pi/2} d\sigma\{[L_\sigma, \cdot]\eta^\sigma + 4i\pi\eta^\sigma\eta^{\prime\sigma}\frac{\delta}{\delta\eta^\sigma}\}$ |

To proceed further, it is convenient to "bosonize" the ghosts, i.e. to find matrix representation as well for the forms. There are standard ways of doing that[4]. The result is equivalent to augmenting spacetime with a time-like periodic dimension $\phi$. For instance the potential 1-form $A$ can be written as $A[X_1, \phi_1; X_2, \phi_2]$. The formulae (6) for matrix multiplication and trace become:

$$(M \cdot N)[X_1, \phi_1; X_3, \phi_3] = \int \delta X_2 \delta\phi_2 M[X_1, \phi_1; X_2, \phi_2] N[X_2, \phi_2; X_3, \phi_3],$$
$$TrM = \int \delta X \delta\phi M[X, \phi; X, \phi]. \tag{16}$$

Once the bosonization is done, all operations are purely matrix operations. By this we mean that multiplication is truly matrix multiplication and integration is just a trace. With these, we can now simply write the Chern-Simons action as

$$I = Tr(A \cdot QA + \frac{2}{3}A \cdot A \cdot A). \tag{17}$$

---

[4]This is explained in many papers, e.g. [4].

By taking the trace, both in $X$ and in $\phi$, we ensure that $I$ is a gauge invariant number.

Not surprisingly, the above set-up looks very much like Witten's theory. The essential difference is that we have written everything in terms of half-strings with $\sigma = 0 \to \pi/2$. The proof that the two formulations are actually equivalent is not all that straight-forward, and has been done to the degree of rigour (or lack of it) consistent with a functional formalism [3]. In particular, the operation of $Q$ on $A$ here is quite different from the usual BRST operator $Q_{BRST}$ on the full-string functional $A_{FS}$, the former by commutation (as is proper with matrices) and the latter by the usual functional derivative. Also implicit in the action are the different normal orderings with respect to $\beta$ and to $\alpha$ respectively. It is in fact quite amazing that the two formulations should be equivalent.

I shall end with two remarks. Firstly, ordinary internal gauge symmetry can be incorporated via the Chan-Paton trace factors. In a way, our matrices can be regarded as gigantic isospin matrices, and the trace works exactly in the same way as the finite-dimensional isospin trace. This suggests that in coupling $n$ strings together to form a gauge invariant quantity in an explicitly dual fashion, trace is all we need. If so, this points to a much more efficient way of calculating $n$-point amplitudes. That this is indeed the case for tree amplitudes has recently been shown [10].

Secondly, all we have done so far is the equivalent of a pure Yang-Mills theory. We may conceivably introduce sources in the form of half-string functionals $\Psi$, and add terms like $\bar{\psi}\,\partial\!\!\!/\,\psi$ of ordinary Yang-Mills theory to the action (17). This may lead to new developments in string theory which could show the way for constructing more realistic physical models.

## Acknowledgements

I am grateful to the Leverhulme Trust for partial financial support.

## References

[1] E. Witten, *Physica Scripta* **T15** (1987) 70.

[2] Chan Hong-Mo and Tsou Sheung Tsun, *Phys. Rev.* **D35** (1987) 2474.

[3] Chan Hong-Mo and Tsou Sheung Tsun, *Phys. Rev.* **D39** (1989) 555.

[4] E. Witten, *Nucl. Phys.* **B268** (1986) 253.

[5] A. Connes, *Publ. Math. I.H.E.S.* **62** (1985) 41.

[6] Chen Ning Yang and R.L. Mills, *Phy. Rev.* **96** (1954) 191.

[7] e.g. Chan Hong-Mo and Tsou Sheung Tsun, *Rutherford Laboratory preprint* RAL-89-052, 1989.

[8] E. Witten, *Nucl. Phys.* **B276** (1986) 291.

[9] J. Bordes, Chan Hong-Mo, Lukas Nellen and Tsou Sheung Tsun, *Rutherford Laboratory preprint* RAL-89-048, 1989.

[10] Chan Hong-Mo, J. Bordes, Tsou Sheung Tsun and Lukas Nellen, *Phys. Rev.* **D**, 1989, *to appear*.

MATHEMATICAL INSTITUTE, 24-29 ST. GILES', OXFORD OX1 3LB, ENGLAND.

# 11

## An Approach to Constructing Rational Conformal Field Theories

*K.S. Narain*

Two-dimensional conformal field theories provide classical solutions to string theory. The simplest ones of these are the rational conformal field theories where the partition functions and the correlation functions can be expressed as finite sums of products of holomorphic and antiholomorphic conformal blocks. Here we shall describe an approach [1] to constructing these finite numbers of conformal blocks. The procedure, as we shall see, is inspired by the Coulomb gas construction of conformal theories [2].

Let us consider $d$ free scalar fields $\phi_i$ (denoted as $d$-dimensional vector $\vec{\phi}$). Let the stress energy be:

$$T_{zz} \equiv T = -\frac{1}{2}(\partial\vec{\phi})^2 + i\vec{\alpha}_0 \cdot \partial^2\vec{\phi},$$

where $\vec{\alpha}_0$ is a constant vector. Then the central charge of the corresponding Virasoro algebra is

$$c = d - 24\alpha_0^2.$$

(We use the normalization $\langle \phi(z_1)\phi(z_2)\rangle = -2\ln(z_1 - z_2)$, for the left sector.)

Let $\{\vec{e}_a\}$ be a set of vectors such that
i) $\vec{e}_a{}^2 - 2\vec{\alpha}_0 \cdot \vec{e}_a = 1$
ii) $e_a^2 = $ rational, i.e. $\exists$ smallest positive integer $k_a$ such that $k_a e_a^2 = $ integer
iii) $2k_a e_a \cdot e_b = $ integer for all pairs $a, b$
iv) $\exists$ positive integers $r_a$ such that $\sum_a r_a \vec{e}_a = 0$.

Condition i) implies that the vertex operators $V_{e_a} =: e^{i\vec{e}_a \cdot \vec{\phi}} :$ are dimension-one operators and hence $\int dz V_{e_a}(z)$ commute with the Virasoro generators. Condition ii) is a requirement for the rational conformal field theories. Note that the dimension of the vertex operator $V_\beta$ is $\Delta_\beta = \beta^2 - 2\vec{\alpha}_0 \cdot \beta = \Delta_{2\vec{\alpha}_0 - \beta}$. Thus we will identify $V_\beta \sim V_{2\alpha_0 - \beta}$. The vertex operators $V_{e_a}$ ( we shall call them screening operators—the terminology will become clearer later), can be used to construct Virasoro primary fields in the $\partial\vec{\phi}$ oscillator modules built on $V_{\vec{\beta}}$. Consider the integral

$$I = \prod_{i=1}^{n_a} \int_{c_i} dz_i V_{e_a}(z_i) V_{2\alpha_0}(0)_{-\beta-n e_a} = P(\partial\phi_i, \partial^2\phi_i \ldots) \cdot V_{2\alpha-\beta}(0),$$

for some closed contours $c_i$ and $P(\partial\phi_1 \ldots)$ is a polynomial of a definite conformal weight. For $I$ to be non-zero one has the following condition

$$2e_a \cdot \beta = (1-n_a)e_a^2 + (m_a - 1), \tag{1}$$

for some positive integer $m_a > 0$. If $\beta$ satisfies the above condition then the polynomial $P$ has the conformal weight $(n_a \cdot m_a)$. Since $\int V_{e_a}$ commutes with the Virasoro algebra, $PV_{2\alpha-\beta}$ is annihilated by all the positive frequency modes of $T$. This means that $PV_{2\alpha_0-\beta}$ is not contained in the Virasoro module $[V_{2\alpha_0-\beta}]$ built on $V_{2\alpha_0-\beta}$. Thus we must subtract this state and its submodule from the oscillator module $\{V_{2\alpha_0-\beta}\}$. However, as far as Virasoro algebra is concerned, one can identify $PV_{2\alpha_0-\beta}$ with $V_{2\alpha_0-\beta-n_a e_a}$ as $V_{e_a}$ commute with the Virasoro algebra. For example, for $\beta = 0$ (i.e. identity) we have the solution $n_a = m_a = 1$ and the $PV_{2\alpha_0} \sim V_{2\alpha_0 - e_a} \equiv V_{2\alpha_0-\beta}$, where $\beta = e_a$. In fact, for a completely degenerate representation, we must require that condition (1) is satisfied for $\beta = e_b$ for each $b$, which is equivalent to the condition iii). Condition iv) arises from the requirements of the correlation function, and we shall describe it later.

Let

$$\Gamma = \{\beta \mid \beta \text{ satisfies Eq. (1) for some positive integers } n_a, m_a\}.$$

We have the symmetry

$$n_a \to n_a + k_a, \quad m_a \to m_a + k_a e_a^2. \tag{2}$$

Thus we can relax the positivity condition. $\Gamma$ is therefore a lattice. Define also

$$\Gamma' = \{\vec{\gamma} \mid \exists \text{ integers } r_a \text{ s.t. } \vec{\gamma} = \sum_a r_a k_a \vec{e}_a\}.$$

Then $\sqrt{2}\Gamma'$ is even integral, and $\Gamma'$ is the dual of $\Gamma$.

The process of subtraction mentioned above can be described by defining a certain "Weyl group" $G$. Define elements $g_a \in G$ as

$$g_a \beta = \beta + n_a e_a.$$

## 11. Constructing Rational CFT

Note that $g_a$ is not well defined on $\Gamma$ due to symmetry (2); however, it is well defined on $\Gamma/\Gamma'$. $g_a : \Gamma/\Gamma' \to \Gamma/\Gamma'$ is a one-to-one onto map. If $\Gamma/\Gamma'$ has $N$ elements then $g_a$ belongs to the permutation group $P_N$. $G$ is defined as the group generated by $g_a$'s. Thus $G \subset P_N$ and hence is a finite group. We must also define a $Z_2$ grading on $G$ to carry out the process of subtraction. Define the grading $c : G \to (+1, -1)$ as

$$c(g_a) = -1,$$

and extend to the entire group $G$. We assume that this extension is well defined (otherwise all the characters would be zero, as we shall see). Besides the "Weyl group" $G$, in general there is also a dimension-preserving group $G_0 : \Gamma \to \Gamma$ (a trivial example is $\beta \to 2\alpha_0 - \beta$). $G_0$ is also well defined on $\Gamma/\Gamma'$ and is a subgroup of $P_N$. Since $G_0$ serves the purpose of identifying states, we define

$$c : G_0 \to +1.$$

Let $G'$ be the smallest group containing $G$ and $G_0$. We will assume that $G_0$ is a normal subgroup of $G'$.

Now we can define the character for each element $\beta \in \Gamma$:

$$\chi_\beta = \frac{1}{\eta d} \sum_{g \in G} c(g) \sum_{P \in \Gamma'} q^{(g\beta - \alpha_0 + P)^2}, \qquad (3)$$

where $q = e^{i 2\pi \tau}$ and $\tau$ is the Teichmüller parameter of the torus. It is easy to see that under modular transformation $\tau \to -1/\tau$

$$\chi_\beta \longrightarrow S_{\beta\beta'} \chi_{\beta'},$$

where

$$S_{\beta\beta'} = \sum_{g \in G} e^{2\pi i (\beta - \alpha_0)(g\beta' - \alpha_0)} c(g).$$

The diagonal modular invariant partition function is

$$Z = \sum_{[\beta] \in \Gamma/\Gamma'} |\chi_\beta|^2,$$

where $[\beta]$ is a representative of $\beta$ in $\Gamma$.

Although the construction of $\chi_\beta$ is, in spirit, that of the usual Coulomb gas formalism, in the absence of the knowledge of chiral algebra that commutes with the screening operators $V_{e_a}$, in general $\chi_\beta$ will not be positive (i.e. in the $q$ expansion for each $\beta$, the coefficients of various powers may not be non-negative). In fact, as our model must be a Virasoro model we have a stronger condition: for $c > 1, \eta \chi_\beta$

must be non-negative except when $\beta$ is such that $\exists g \in G$ such that $[g\beta] = 0$ and in the latter case

$$n\chi_\beta = q^{-(c-1)/24}(1 - q + \sum_{N=2}^{q} a_n q^n),$$

with integers $a_n \geq 0$. Of course, for $c < 1$, we have only the requirement that $\chi_\beta$ be positive.

The positivity condition turns out to be very severe, and in fact all the examples we have found so far satisfying this condition turn out to be one of the GKO coset constructions. Indeed, one of the most important open problems in this approach is to encode the positivity condition directly on the choice of the screening charges $\vec{e}_a$.

To give a non-trivial example, consider $d = 2$ and

$$e_{1,\pm} = \begin{pmatrix} (k+2)\alpha_0/2 \\ \pm\sqrt{3}/4 \end{pmatrix}, \quad e_{2,\pm} = 2\alpha_0 - e_\pm, \quad \alpha_0 = \begin{pmatrix} 1/\sqrt{k^2-4} \\ 0 \end{pmatrix}$$

for $k$ odd. One can proceed with the construction above and one obtains at the end characters of the spin $4/3$ parafermionic series $c = (2 - \frac{24}{(k-2)(k+2)})$ for odd $k$. (Even $k$ can be obtained with slight modification of the screening charges.) The first model in this series (for $k = 5$) happens to be the Virasoro model $c = 6/7$. One of the striking features of this series is that there is a field $V_\beta$, $\beta = (0, \sqrt{1/3})$ with weight $1/3$; however, in the character $\chi_\beta$, there is a state of same weight with negative sign $g\beta = \begin{pmatrix} 7\alpha_0/2 \\ 1/\sqrt{12} \end{pmatrix}$ with $c(g) = -1$, so that $q^{1/3-c/24}$ cancels out and the character $\chi_\beta$ starts with $q^{4/3-c/24}$. Indeed, as is known in the spin-$4/3$ parafermionic series studied by Fateev and Zamolodchikov [3] the dimension-$1/3$ state is a null state of the parafermionic algebra and one starts with a dimension-$4/3$ field. In our formalism the disappearance of the dimension-$1/3$ field occurs as stated above. However, the important point is that even though the vertex operator $V_\beta$ and the $V_{g\beta}$ of dimension $1/3$ disappear, their first-level secondaries would contribute to the dimension-$4/3$ state. In other words, the dimension-$4/3$ primary field would be a linear combination of the first-level secondaries of $V_\beta$ and $V_{g\beta}$ and the dimension-$4/3$ vertex operator $V_{\beta'}$ ($\beta' = \begin{pmatrix} 0 \\ 2/\sqrt{3} \end{pmatrix}$). We shall describe this precise combination below in the context of construction of correlation functions.

Let us consider the four-point function

$$\prod_i \int dz_i \langle V_\beta(0) V_\alpha(z) V_\alpha(1) V_{e_{a_i}}(z_i) V_{2\alpha_0-\beta}(\infty) \rangle \equiv I(z), \quad (4)$$

## 11. Constructing Rational CFT

where by conservation of charges (including the background charge $-2\alpha_0$) we have
$$2\alpha + \sum_i e_{a_i} = 0.$$

For simplicity, let us consider the case where $\alpha$ is such that $V_\alpha$ does not disappear in the fashion described above. Suppose we are interested in constructing the four-point function for $V_\beta$ corresponding to the dimension-4/3 field. How can one take the correct combination? To answer this, it is best to go to the two-point function on the torus

$$\prod_i \int dz_i \langle V_\alpha(x_1) V_\alpha(x_2) V_{e_{a_i}}(z_i) \rangle_\beta \equiv I_t(x_1 - x_2),$$

where $\langle \ \rangle_\beta$ denotes the two-point function evaluated in the $\beta$-character, i.e. the trace is taken over the $[\beta]$ representation. Constructions of correlation functions on the torus have been described in [4]

$$I_t(x_1 - x_2) \propto (\prod_i \int_{c_i} dz_i) \theta(x_1 - x_2)^{2\alpha^2} \prod_i \theta(x_1 - x_2)^{2\alpha \cdot e_{a_i}}$$
$$\prod_{i<j} \theta(z_i - z_j)^{2e_{a_i} \cdot e_{a_j}} \times \Gamma_\beta(x_1, x_2, z_i),$$

where

$$\Gamma_\beta(x_1, x_2, z_i) = \sum_{g \in G'} c(g) \sum_{p \in \Gamma'} q^{(g\beta - \alpha_0 + P)^2}$$
$$\exp(4\pi i (g\beta - \alpha_0 + P) \cdot (\alpha(x_1 + x_2)) + \sum_i e_{a_i} z_i). \tag{5}$$

Note that the $\theta$-functions are the usual correlation functions of vertex operators and $\Gamma_\beta$ is just the character with internal momenta coupling to the external sources. The contours $c_i$ in the $t$-channel are between $(x_1, x_2)$ or around the $b$-cycle, and in the $s$-channel the two independent contours are (in the annulus picture of the torus) between 0 and $x_1$ and between $x_2$ and $\infty$. To obtain conformal blocks (in the $s$-channel) that transform correctly under $b$-cycle monodromy the contours $c_i$ will in general depend on the element $g \in G'$. Indeed under $x_1 \to x_1 + \tau$ if $e_{a_i}$ screening contours go around $x_1$, then these contours are also pulled along so that $z_i \mapsto z_i + \tau$. From eq. (5) this means that
$$g\beta \longrightarrow g\beta + \alpha + \sum_i e_{a_i}.$$

This is, in general, not equal to $g(\beta + \alpha + \sum_i e_{a_i})$. Thus for different $g \in G'$ one must choose different contours $c_i$ in the $s$-channel. One can show that such a choice always exists if condition (iv) is satisfied.

Returning to the four-point functions on the sphere, we can take the degeneration limit $q \to 0$, $x_i \to \ln x_i/2\pi i$, $z_i \to \ln z_i/2\pi i$ in eq. (5) to obtain the four-point function (4). This procedure gives the correct combination required for the $V_\beta$ state. For example, in the case of the spin-4/3 parafermionic series for $c = 6/7$ and $\beta = \begin{pmatrix} 0 \\ 1/\sqrt{3} \end{pmatrix}$ corresponding to the dimension-1/3 field (which disappears from the characters), upon degeneration, the leading term is of the order of $q^{4/3-c/24}$ and it gets contributions from several terms: expansion of $\theta$-functions with $\beta = \begin{pmatrix} 0 \\ 1/\sqrt{3} \end{pmatrix}$ corresponding to the first oscillator secondary on $V_\beta$ as well as the vertex operator $V_{\beta'}$, $\beta' = \begin{pmatrix} 0 \\ 2/\sqrt{3} \end{pmatrix}$ corresponding to the dimension-4/3 field. We have checked that these terms reproduce the correct four-point function. This exercise, while providing a prescription to compute four-point functions for this generalized situation, is also in fact a non-trivial check on the prescription to compute two-point functions on the torus, as it involves expansions to higher order in $q$.

In conclusion, we have described here a method to construct characters and correlation functions that form finite dimensional representations of modular and monodromy groups. These representations are based on the choice of the screening charges; however, one important open problem in this approach is that of understanding better the positivity condition. If the latter condition is satisfied, one would then have a well-defined conformal field theory. One could then in principle obtain some information on the chiral algebra or at least the spectrum-generating algebra [5].

One can also generalize the above construction to rational Lorentzian lattices. Indeed, one already knows of such examples: $N = 2$ minimal models ($c < 3$), $SU(2)$ parafermions and $SU(2)$ WZW models [6]. In these cases one of the bosons appears with imaginary momentum. These characters can be obtained by following the procedure outlined above, the only difference being that one must restrict the $\Gamma'$ lattice sum inside the light cone (i.e. positive dimensions). Thus the character becomes [we specialize here to $d = 2$ and signature $(+1, -1)$]

$$\chi_\beta = \sum_{g \in (G'/\tilde{G}_0)} c(g) \sum_{P \in \Gamma' \cap L_+/\tilde{G}_0} q^{(g\beta - \alpha_0 + P)^2}, \qquad (6)$$

where $L_+$ is the positive cone with positive dimensions and $\tilde{G}_0$ is the dimension-preserving automorphism $\subset$ connected component of the Lorentz group (this is because $\tilde{G}_0$ is of infinite order, so in order to

## 11. Constructing Rational CFT

get a finite answer, one must divide by $\tilde{G}_0$). In general, the characters $\chi_\beta$ will *not* form a finite dimensional representation of the modular group due to the restricted lattice sums $\Gamma' \cap L_+$. However one can show that if $G'$ contains an element $\tilde{g}$ such that

$$\tilde{g}[\beta] = [2\alpha_0 + \beta - \frac{2\alpha_0(\alpha_0 \cdot \beta)}{\alpha_0^2}]$$

on $\Gamma/\Gamma'$ and $c(\tilde{g}) = -1$ then Eq. (6) can be rewritten as

$$\chi_\beta = \sum_{g \in (G'/\tilde{G}_0)/Z_2} c(g) ( \sum_{P \in \Gamma' \cap L_+/\tilde{G}_0} - \sum_{P \in \Gamma' \cap L_-/\tilde{G}_0} ) q^{(g\beta - \alpha_0 + P)^2}, \qquad (7)$$

where $Z_2$ is the subgroup $(1, \tilde{g}) \subset G'$. Such a subgroup $Z_2$ exists for all the known examples involving Lorentzian lattices. The remarkable property of this particular combination

$$( \sum_{\Gamma' \cap L_+} - \sum_{\Gamma' \cap L_-} )$$

is that this combination is modular invariant even though the lattice sums are restricted. This is a consequence of the following identity

$$(\int_{L_+} d^2x - \int_{L_-} d^2x) e^{-i\pi(x^2/2\tau) + 2\pi i P \cdot X} = \begin{cases} +q^{P^2} & \text{if } P \in L_+ \\ -q^{P^2} & \text{if } P \in L_- \\ 0 & \text{otherwise.} \end{cases}$$

Thus

$$( \sum_{P \in \Gamma' \cap L_+} - \sum_{P \in \Gamma' \cap L_-} ) q^{(P+\beta)^2}$$

$$= \sum_{P \in \Gamma'} (\int_{L_+} d^2x - \int_{L_-} d^2x) e^{-i\pi(x^2/2\tau) + 2\pi i (P+\beta) \cdot X}$$

$$= (\int_{L_+} d^2x - \int_{L_-} d^2x) e^{-i\pi(x^2/2\tau)} \sum_{P \in \Gamma'^* = 2\Gamma} \delta^2(x - P) e^{2\pi i \beta \cdot X}$$

$$= \sum_{P \in \Gamma \cap L_+} - \sum_{P \in \Gamma \cap L_-} ) e^{(-2\pi i/\tau) P^2 + 2\pi i \beta \cdot P}.$$

Using these methods one can also construct the correlation function on the sphere and the torus for $SU(2)$ parafermions [7] and $N = 2$ superconformal minimal models [8].

It should be possible to carry out a similar analysis for higher number of bosons ($d > 2$). For $SU(3)$ and in the general $SU(N)$ WZW model, there exists a Coulomb gas representation [6] in terms of $R$ (=rank) scalar fields and $(D - R)/2$ pairs of commuting ghost systems $(\beta, \gamma)$ with conformal weights $(1, 0)$, where $D$ is the dimension of the group. By bosonizing $(\beta, \gamma)$ systems, one would therefore get $(D - R)/2$ pairs of two-dimensional lattices with signatures $(+1, -1)$. In view of this, it is quite possible that all the GKO coset models may have a Coulomb gas description. The question is whether the Coulomb gas description can provide also unitary models other that GKO cosets or whether the latter exhaust all the unitary rational conformal models (for $c < 24$).

## Acknowledgements

This talk is based on a work in collaboration with M. Caselle [1]

## References

[1] M. Caselle and K.S. Narain, *Nucl. Phys.* **B323** (1989) 673.

[2] V. Dotsenko, V.A. Fateev, *Nucl. Phys.* **B240** [**FS12**] (1984) 312, **B251** (1985) 691;
V.A. Fateev and A.B. Zamolodchikov, *Nucl. Phys.* **B280** (1987) 644;
V.A. Fateev and S.L. Lykyanov, *Int. Jour. Mod. Phys.* **A3** (1988) 507.

[3] V.A. Fateev, A.B. Zamolodchikov, *Theor. Math. Fiz.* **71** (1987) 163.

[4] G. Felder, *Zurich preprint* 88-0618 (1988);
T. Jarayaman and K.S. Narain, *CERN preprint*, 1988.
J. Bagger, D. Nemeschansky and J.B. Zuber, *Harvard preprint* 1988;
Also in hyperelliptic formalism:
Al.B. Zamolodchikov, *Nucl. Phys.* **B316** (1989) 573;
C. Crnkovic, G.M. Sotkov and M. Stanishkov, *ICTP preprint*, 1988.

[5] S.D. Mathur, S. Mukhi and A. Sen, *TIFR preprints* TIFR/TH/88-32 and 88-50

[6] M. Wakimoto, *Comm. Math. Phys.* **104** (1986) 605;
A.B. Zamolodchikov, Montreal Lectures, 1988, unpublished;
A. Gerasimov, A. Marshakov, A. Morozov, M. Olshanetsky and S. Satashivili, *ITEP preprint*, 1989;
D. Nemeschansky, Feigin-Fuchs representation of string functions, *University of Southern Carolina preprint* USC-89/012.

[7] T. Jarayaman, K.S. Narain and M.H. Sarmadi, *in preparation*.

[8] T. Jarayaman, M.A. Namazie, K.S. Narain, C. Nunez and M.H. Sarmadi, *in preparation*.

THEORY DIVISION, CERN, GENEVA, SWITZERLAND.

# 12

## A Universal Link Invariant

R.J. Lawrence

### 1. Review of standard theory.

Recall that the braid group $B_n$ is the fundamental group of the configuration space of $n$ points in the plane. It has a presentation given by generators $\sigma_1, \ldots, \sigma_{n-1}$ with relations:

$$\begin{aligned}\sigma_i \sigma_{i+1} \sigma_i &= \sigma_{i+1} \sigma_i \sigma_{i+1} & &\text{for } i = 1, 2, \ldots, n-2 \\ \sigma_i \sigma_j &= \sigma_j \sigma_i & &\text{for } |i - j| > 1\end{aligned}$$

The Iwahori-Hecke algebra $H_n(q)$ is the quotient of $B_n$ given by imposing the relation:

$$(\sigma_i - 1)(\sigma_i + q) = 0$$

for all $i = 1, 2, \ldots, n-1$. Note that $H_n(q)$ reduces to the symmetric group, $S_n$, for $q = 1$.

The operation of closure, in which the ends of the braid are canonically closed, as indicated in the diagram below, gives rise to a link in $S^3$ — that is, a union of disjoint non-self-intersecting closed curves. By Alexander's theorem [1], this procedure, when applied to all braids for all $n$, will produce all links in $S^3$, up to isotopy. By Markov's theorem [2], two braids, in possibly different sized braid groups correspond to isotopic links, if and only if, one can pass from one to the other by a sequence of moves, each being one of the types below:

$$\begin{aligned}\alpha &\longrightarrow \beta \alpha \beta^{-1} & &\text{for } \alpha, \beta \in B_r \\ \alpha &\longleftrightarrow \alpha \sigma_r^{\pm 1} & &\text{for } \alpha \in B_r \hookrightarrow B_{r+1}\end{aligned}$$

Thus an invariant of links is obtained whenever representations $\pi_n$ of the braid group $B_n$ are given for each $n$, together with traces $\text{tr}_n$ which satisfy the *Markov condition*:

$$\text{tr}_{n+1}\left(\pi_{n+1}(\gamma \sigma_n^{\pm 1})\right) = \text{tr}_n\left(\pi_n(\gamma)\right).$$

This invariant is $\text{tr}_n(\pi_n(\gamma))$ for the link obtained by closing $\gamma$.

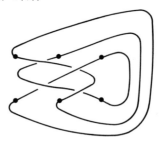

The above results give rise to one of the main classes of methods for obtaining link invariants. Suppose $L$ is a link, expressed as the closure of a braid $\gamma \in B_n$: this is denoted by $\hat\gamma \sim L$. One of the ways of defining the Jones polynomial is to project $\gamma$ to $\pi_n(\gamma) \in H_n(q)$ and then obtain an invariant as a certain normalisation of:

$$\text{tr}(\pi_n(\gamma));$$

where tr is the *Ocneanu trace* defined on $H_\infty(q) = \amalg H_n(q)$. by:

$$\begin{aligned} \text{tr}(1) &= 1 \\ \text{tr}(g\sigma_n) &= z\text{tr}(g) \end{aligned}$$

whenever g is a word in $\sigma_1,\ldots,\sigma_{n-1}$. We then obtain the invariant

$$X_L = \left(\frac{\lambda q - 1}{\sqrt{\lambda}(1-q)}\right)^{n-1} \left(\sqrt{\lambda}\right)^e \text{tr}(\pi(\gamma))$$

where $e$ is the exponent sum of $\gamma$ as a word in $\{\sigma_i\}$, and $\lambda, q, z$ are related in a suitable way (see [3] for more details). Here $X_L$ is the two variable Jones polynomial.

Another way of obtaining a representation of $B_n$ is to use a representation of the form:

$$\begin{aligned} \pi_n : B_n &\longrightarrow \text{End}(V^{\otimes n}) \\ \sigma_i &\longmapsto \hat{R}_{i\,i+1} \end{aligned}$$

where $\hat R \in \text{End}(V \otimes V)$ and $V$ is a finite dimensional vector space. Here $\hat R_{i\,i+1}$ denotes the action of $\hat R$ on the $i^{th}, (i+1)^{th}$ factors of $V^{\otimes n}$, and the identity on the rest. This defines a representation of $B_n$ so long as $\hat R$ satisfies the *braid relation*:

$$\hat R_{12} \hat R_{23} \hat R_{12} = \hat R_{23} \hat R_{12} \hat R_{23}$$

as elements of $\text{End}(V \otimes V \otimes V)$. One can now obtain an invariant of links so long as a suitable family of traces can be found satisfying

## 12. A Universal Link Invariant

the Markov property. This can be done if $\hat{R}$ can be extended to a so called *enhanced Yang-Baxter operator*: see [4] for more details.

Turaev also showed how the invariant so obtained can be put in the form of a state model on the diagram $\mathcal{D}$ obtained as a two-dimensional projection of the link $L$. We assume that $L$ is initially oriented; then $\mathcal{D}$ has oriented edges, and its vertices correspond to crossings in $L$. The state model defined by Turaev has *states* $\sigma$:

$$\sigma : \mathrm{Edg}(\mathcal{D}) \to \{1, 2, \ldots, m\}$$

where $\{1, 2, \ldots, m\}$ is an indexing set for a basis of $V$. For a set $\sigma$, he then assigns to a vertex $v$ at which four edges $a, b, c, d$ come together, a local weight $w_v(\sigma)$ according to the relative orientations of the edges, as shown below:

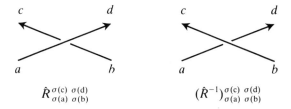

The invariant is then computed as:

$$\alpha^{-w(\mathcal{D})} \beta^{-n} \sum_\sigma \left( \prod_v w_v(\sigma) \right) \left( \prod_i \mu_i \right)$$

where $w(\mathcal{D})$ is the *writhe* of the diagram. Here $\alpha, \beta \in \mathbf{C}$ and $\mu : V \longrightarrow V$ are the additional objects required to obtain an enhanced Yang-Baxter operator from $\hat{R}$. Again, we refer the reader to [4] for more details: we here only outline the shape of the link invariant.

### 2. Quantum Groups

Suppose $\mathfrak{g}$ is a Lie group. It can be shown (see [5], [6]) that there corresponds a *quantum group* $U_h\mathfrak{g}$, together with a natural element:

$$R \in U_h\mathfrak{g} \otimes U_h\mathfrak{g}$$

satisfying the (constant) quantum Yang-Baxter equation (QYBE):

$$R_{12} R_{13} R_{23} = R_{23} R_{13} R_{12}.$$

Here $R_{ij}$ denotes the element of $(U_h\mathfrak{g})^{\otimes 3}$ defined by $R$ in the $i^{th}, j^{th}$ factors and $1 \in U_h\mathfrak{g}$ in the third. A quantum group should be thought of as a quantum universal enveloping algebra which is a deformation of $\mathfrak{g}$.

A representation $\rho$ of $U_h\mathfrak{g}$ gives rise to a solution $\rho^*(R)$ of the QYBE in $\text{End}(V \otimes V)$. If we define $P \in \text{End}(V \otimes V)$ to be the map transposing the factors, then it can be easily seen that:

$$\hat{R} = P \circ \rho^*(R)$$

is a solution of the braid relation on $V \otimes V$, and can be used to compute invariants of links as indicated in §1. Thus from any Lie group and representation of the associated quantum group we can obtain a link invariant — what we call the generalised Jones polynomial. There are some extra complications that come in here, involving enhanced Yang-Baxter operators, but we have no space to discuss this here. However the general procedure for producing invariants is as we specified above.

The representation $\rho$ enters into the invariant in two ways: to obtain a solution of the braid relation from $R$; and to supply a trace. Since $P$ does not correspond to any element of $U_h\mathfrak{g} \otimes U_h\mathfrak{g}$, then although $R$ is expressible universally in $U_h\mathfrak{g} \otimes U_h\mathfrak{g}$, $\hat{R}$ is not so expressible. Thus the natural method for producing a representation of $B_n$, namely by mapping it into $(U_h\mathfrak{g})^{\otimes n}$, cannot be used. Instead we define the twisted product:

$$(U_h\mathfrak{g})^{\otimes n} \rtimes S_n$$

to be the semi-direct product with $S_n$ acting on $(U_h\mathfrak{g})^{\otimes n}$ by permuting the factors. We then define:

$$\begin{aligned} \pi_n : B_n &\longrightarrow (U_h\mathfrak{g})^{\otimes n} \rtimes S_n \\ \sigma_i &\longmapsto R_{i\,i+1} \cdot (i\,i+1) \end{aligned}$$

It is easily checked that $\pi_n$ is well defined as a homomorphism. Suppose that $L$ is a link, expressed as the closure of a braid $\gamma \in B_n$. Then:

$$\pi_n(\gamma) = g \cdot \sigma$$

where $\sigma \in S_n, g \in (U_h\mathfrak{g})^{\otimes n}$. In fact $\sigma$ is the natural image of $\gamma$ in the quotient $S_n$ of $B_n$, and thus the number of cycles in the disjoint cycle decomposition of $\sigma$ is precisely the number of components, $k$ say, of $L$.

To obtain an invariant of links, it is now necessary to use a suitable trace, for each $n$, such that the Markov property is satisfied. This can be done, and one then obtains an invariant in $(U_h\mathfrak{g}/I)^{\otimes k}$ where $I$ is a subspace of $U_h\mathfrak{g}$ which is spanned by "generalised commutators." This is the *universal link invariant*. For more details see [7]. This invariant has one factor for each component of the link — or, equivalently, for each cycle in the disjoint cycle decomposition of $\sigma$. Corresponding to a cycle $(a_1, \ldots, a_l)$ in $\sigma$, the term in the invariant is:

$$\sum X^{(l-1)}(g_{a_l} \cdots g_{a_1})$$

## 12. A Universal Link Invariant

where $g \in (U_h\mathfrak{g})^{\otimes n}$ is written as $\sum g_1 \otimes \cdots \otimes g_n$. Here $X^{(l-1)}$ denotes the composition of $(l-1)$ copies of a linear map $X$ on $U_h\mathfrak{g}$.

### 3. State model formulation

Let $\mathcal{J}$ be an indexing set for a basis of $U_h\mathfrak{g}$. Then we can expand $R, R^{-1}$, as:

$$\sum_{i \in \mathcal{J}} a_i \otimes b_i, \quad \sum_{i \in \mathcal{J}} b'_i \otimes a'_i$$

respectively. We now define a state model formulation on a diagram $\mathcal{D}$ of the oriented link $L$ as follows. We assume that $\mathcal{D}$ is the diagram of $L$, drawn as the closure of a braid. Then a *state* of the model is a map:

$$\sigma : \text{Vert}(\mathcal{D}) \longrightarrow \mathcal{J}.$$

We fix a point on $L$, and then go around $L$ once, coming back to the starting point. For a state $\sigma$, as we encounter a crossing we associate a local weight $w_v(\sigma)$ in accordance with the diagrams below. Here $v \in \text{Vert}(\mathcal{D})$ is the vertex corresponding to the crossing concerned; and in the diagrams, a double arrow indicates the path followed, while, a single arrow indicates the path crossed.

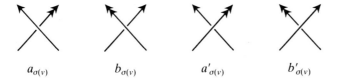

$$a_{\sigma(v)} \qquad b_{\sigma(v)} \qquad a'_{\sigma(v)} \qquad b'_{\sigma(v)}$$

We can now write the invariant as:

$$X^{(l-1)} \left( \sum_\sigma \prod_v w_v(\sigma) \right)$$

where the product over vertices $v$ must be taken in the order in which they are traversed, starting at some point on a component. For a link with more than one component, one must compute the products of local weights around each component separately, and then take their tensor product before summing over states. We then apply:

$$X^{(l_1-1)} \otimes \cdots \otimes X^{(l_k-1)}$$

to the resulting sum, obtaining an element of $(U_h\mathfrak{g})^{\otimes k}$ which must be projected into $(U_h\mathfrak{g}/I)^{\otimes k}$ to obtain the invariant.

### 4. Some concluding remarks

1. In the last section we assumed that the components of the link are coloured. When the components of the link are uncoloured

the invariant obtained is the symmetric version of that described in the last section, with respect to permutation of the $k$ factors of the invariant.

2. An invariant has been constructed for *any* quantum group together with a solution of the YBE on it — rather than just those quantum groups which are obtained by quantising Lie algebras. The invariants computed from $U_h\mathfrak{g}$, for fixed $g$, and any $R$ satisfying the QYBE on $U_h\mathfrak{g} \otimes U_h\mathfrak{g}$ contain precisely the same information as all the generalised Jones polynomials evaluated using all representations of $\mathfrak{g}$.

3. There are clear similarities between our state model formulation of the universal link invariant and Jones' formula (see [3]§1, p.352). Thus:

$$(\sqrt{\lambda})^\epsilon \pi(\gamma) \longleftrightarrow \sum_\sigma \prod_v w_v(\sigma)$$

$$\frac{\lambda q - 1}{\sqrt{\lambda}(1-q)} \longleftrightarrow X$$

However the trace in the Jones polynomial does not correspond to anything in our formula: it corresponds to choosing a representation of $U_h\mathfrak{g}$.

## References

[1] J.W. Alexander, Topological invariants of knots and links, *Trans. Amer. Math. Soc.* **20**(1923)275-306

[2] A.A. Markov, Über die freie Aquivalenz geschloßener Zöpfe, *Recueil Mathematique Moscou* **1**(1935)73-78

[3] V.F.R. Jones, Hecke algebra representations of braid groups and link polynomials, *Ann. Math.* **126**(1987)335-388

[4] V.G. Turaev, The Yang-Baxter equation and invariants of links, *Invent. Math.* **92**(1988)527-553

[5] V.G. Drinfel'd, Quantum Groups, in *Proc. Intl. Congress Math.*, Berkeley (1986)789-820

[6] M. Jimbo, A $q$-difference analogue of $U(q)$ and the Yang-Baxter equation, *Lett. Math. Phys.* **10**(1985)63-69

[7] R.J. Lawrence, A universal link invariant using quantum groups, *Oxford University Preprint*, 1988

MATHEMATICAL INSTITUTE, 24-29 ST. GILES', OXFORD OX1 3LB, ENGLAND.

# 13

## Strings and Quantum Gravity

*H.J. de Vega and N. Sánchez*

One of the main challenges in theoretical physics today is the unification of all interactions including gravity. At present, string theories appear as the most promising candidates to achieve such a unification. However, gravity has not completely been incorporated in string theory, many technical and conceptual problems remain and a full quantum theory of gravity is still non-existent. Till now, strings are more frequently formulated in *flat* space-time. Gravity appears through massless spin two particles (graviton). One disposes only of partial results for strings in curved backgrounds [1]–[3]; these mainly concern the problem of consistency (validity of quantum conformal invariance) through the vanishing of the beta-functions. The non-linear quantum string dynamics in curved space-geometry (background field method) where the field propagator is essentially taken as the flat space Feynman propagator. Clearly, such approximations are useless for the computation of physical quantities (*final parts*) as the mass operator, scattering amplitudes and critical dimension in strong curvature regimes where the quantum gravity effects are important. Our aim is to properly understand strings in the context of quantum gravity. A first step in this program was the investigation by the present authors of string quantization in Rindler space-time (with application to Schwarzschild geometry) and the Hawking-Unruh effect in string theory (see [4] and [5]). On the other hand, we have developed a new approach [6] to string quantization in curved space-times taking into account strong curvature effects of the geometry. In our method, we start from an exact solution of the string equations in a given curved metric and then we construct a perturbative series around this solution. The exact (zero-order) solution fully takes into account gravitational effects including those of the singularities in the geometry. Then, first and second order quantum fluctuations around this solution are computed. (This is different from the background field method where only flat space propagators are used). In our method, the quantum fluctuations lead to physical magnitudes which fully take into account the effect of the space-time geometry. We have applied this method to quantize strings in de Sitter space and found the mass spectrum, critical dimension and Regge trajectories in such space [6]. Here we apply the formalism to describe the quantum string dynamics in the Schwarzschild geometry and describe the effects of the scattering and interaction between the string and the black hole. We analyse the equations of motion and constraints both in the Schwarzschild and Kruskal manifolds and their asymptotic behaviours. The angular equation does not decouple from the equations for the radial ($R$) and temporal ($X^\circ$) components. The free equation is obtained only if $R$ happens to be large ($R \gg R_S$) but not near the horizon. ($R_S$ is the Schwarzschild radius). A convenient parametrisation is introduced to deal with the string in cartesian type coordinates instead of spherical ones. As an exact string solution in $D$ dimensional Schwarzschild space-time we take the centre of mass motion (c.m.) of the string. This is explicitly solved by quadratures. The radial c.m. goes to $\infty$ when $\tau \to \infty$ where $\tau$ labels the proper time of the c.m. trajectory. We recognize an effective potential and analyse their properties. It has an attractive singularity at the origin whereas the attractive newtonian tail is absent for $D > 5$ (and non-zero angular momentum). We find the total absorption cross section which generates the $D = 4$ one. We find the first and second order quantum fluctuations around this (c.m.) solution and find the energy momentum tensor including the second order. We describe the scattering of a (first quantized) string by the black hole. For large $\tau$ (large $R$) the equations of motion are free equations. We define an in-basis of solutions in which we expand first and second order quantum fluctuations and define left and right oscillation modes of the string in the asymptotically flat regions of the space-time. We have two alternative definitions depending on whether we take ingoing ($\tau \to -\infty$) or outgoing ($\tau \to +\infty$) regions. The "in" and "out" coefficient modes are related by a linear transformation describing transitions between the internal oscillatory modes of the string as a consequence of the scattering by the black hole. We find two main effects: (i) a change of *polarisation* of the modes without changing their left or right character and (ii) a *mixing* of the particle and antiparticle modes *changing* at the same time their left or right character. That is, (i) if in the ingoing state the string is in an excited mode with a given polarisation, then in the outgoing state there will be non-zero amplitudes for modes polarized in any direction and (ii) an amplitude for an *antiparticle* mode polarized in the same direction but with the right (or left) character reversed. We study the $L_n$ (conformal) generators and the constraints. An easy way to deal with the gauge invariance associated with the conformal invariance on the world sheet is to take the light cone gauge in the ingoing

## 13. Strings and Quantum Gravity

direction. Time evolution from $\tau = -\infty$ to $\tau = +\infty$ conserves the physical or gauge character of the modes. The independent physical excitations are those associated with the transverse modes. We solve for the second order fluctuations in a mode representation and get the energy momentum tensor. Since $T_{\pm\pm}$ is conserved, the $L_n$ generators can be computed in terms of the ingoing basis ($\tau = -\infty$) or alternatively of the outgoing basis ($\tau = +\infty$). The conservation of $L_n$ yields $L_n(\alpha^{in}) = L_n(\alpha^{out}) + \triangle L_n$, where $L_n$ is bilinear in the $\alpha^{in}$'s and describes excitations between internal (particle) states of the string due to the scattering by the black hole. The $L_n$ conservation provides a set of quadratic conditions on the linear transformation between the in and out states. We find the mass spectrum from the $L_0 \approx 0$ constraint, which is formally the same as in flat space-time. This is a consequence of the asymptotically flat character of the space-time and of the absence of bound states for $D > 4$ (if bound states existed, they would appear in the (mass)$^2$ operator besides the usual flat space spectrum). The critical dimension at which massless spin-two states appear is $D = 26$. We obtain explicit expressions for the deflection function, elastic and inelastic cross sections of the string by the black hole. Pair creation takes place (and in all the modes), as a consequence of the composite structure (oscillator modes) of the string. The $S$-matrix amplitudes including the pair creation rate are found at first order in $\sqrt{\alpha'}/R_S$ ($\sqrt{\alpha'}$ = Planck length and $R_S$ = Schwarzschild radius). Explicit computations are made in the weak field expansion in powers of $(R_S/b)^{D-3}$ ($b$ = impact parameter of the string centre of mass). The quantum string corrections to the gravitational analogue of Rutherford's scattering are computed (these are of order $\alpha'^2$). The pair creation or radiation amplitude we find here is of order $\alpha'$; it is *non-thermal* and of *different origin* than Hawking radiation. Detailed calculations are given in [7] and [8]. As it is well known, the Schwarzschild matric is Ricci flat. Therefore, the $\beta$-function vanishes at the leading order both in field theory loops and string loops. Taking into account non-leading terms of $\beta$ leads to correcting the black-hole metric by terms of order $(\ell_{Pl}/R_S)^2$ and higher. One sees that these corrections are of the same order as the higher corrections in our string quantization scheme. Therefore it is consistent to use the classical Schwarzschild metric to quantize strings at the leading order, as we did in the present paper. One must take into account corrections to the metric in order to compute the subleading quantum corrections.

We have also studied point particle field and quantum string scattering in the Aichelburg–Sexl geometry. This is the shock wave metric created by ultra-high energy particles. The field and the string equations of motion and constraints can be exactly solved in this geometry. These results have been reported in [9] and [10].

## References

[1] C. Lovelace, *Phys. Lett.* **135B** (1984)75; *Nucl. Phys.* **B273** (1986) 413.

[2] E. S. Fradkin and A. A. Tseylin, *Nucl. Phys.* **B261** (1985)1.

[3] C. G. Callan, E. J. Martinec and M. J. Perry, *Nucl. Phys.* **B262** (1985)593;

S. Jain, R Shankar and S. R. Wadia, *Phys. Rev.* **D32** (1985)2713;

D. Nemeschansky and S. Yankielowicz, *Phys. Rev. Lett.* **54** (1985)620;

J. Maharana and G. Veneziano, *Nucl. Phys.* **B283** (1987)126;

C. Curci and G. Paffuti, *Nucl. Phys.* **B286** (1987)399;

C. M. Hull and P. K. Townsend, *DAMPT Preprint*, 1986

A. Sen, *SLAC-PUB* preprint 4136, 1986

S. P. de Alvis, *Phys. Rev.* **D34** (1986)3760

C. G. Callan, I. R. Klebanov and M. J. Perry, *Nucl. Phys.* **B278** (1986)78; **B272** (1986)647;

W. Fischler and L. Susskind, *Phys. Lett.* **171B** (1986)383; *Phys. Lett.* **173B** (1986)262;

C. G. Callan, C. Lovelace, C. R. Nappi and S. A. Yost, *Nucl. Phys.* **B288**52;

[4] H. J. de Vega and N. Sánchez, *Nucl. Phys.* **B299** (1988)818;

[5] N. Sánchez, *Phys. Lett.* **B195** (1987)160.

[6] H. J. de Vega and N. Sánchez, *Phys. Lett.* **B197** (1987)320.

[7] H. J. de Vega and N. Sánchez, The scattering of strings by a black hole, to appear in *Nucl. Phys.* **B**.

[8] H. J. de Vega and N. Sánchez, Quantum Dynamics of strings in black Hole space-time, to appear in *Nucl. Phys.* **B**.

[9] H. J. de Vega and N. Sánchez, *LPTHE/DEMIRM* preprint 88-11, to appear in *Nucl. Phys.* **B**.

[10] H. J. de Vega and N. Sánchez, *LPTHE/DEMIRM* preprint 88-25, to appear in *Nucl. Phys.* **B**.

LPTHE, Université Pierre et Marie Curie, Paris, France.

DEMIRM, Observatoire de Paris, Section de Meudon, France.

# 14

## Quantum Group generalizations of String Theory

*H.J. de Vega and N. Sánchez*

Within the perspective to understand physics at Planck scale, we give a quantum group generalization of string theory. A deformation of the Heisenberg commutation relations defines a one parameter family of string models. A hamiltonian approach is developed. We find the effect of the quantum group deformation on the Virasoro algebra. We construct the $N$-point quantum group deformed string amplitudes for closed and open strings and analyze their properties. We find the spectrum and high energy behaviour. The quantum group deformation removes the usual equal point singularities of propagators and shifts the intercept to $s = 0$. In spite of the fact that the theory exhibits Regge trajectories, the behaviour for $s \to \infty$ is of Born type.

In the context of quantum gravity, string theory plays a privileged role: it gives a uv finite theory including the graviton. In the absence of experimental guidance it is worthwhile to explore new mathematical structures connected with string theory, infinite dimensional algebras and beyond. The study of integrable theories and their associated Yang–Baxter (YB) algebras [1] suggested deformations of the universal enveloping Lie algebras [2]. This is now called a "quantum group". More recently, vertex representations of these algebras have been given [3,4]. They involve operators fulfilling Heisenberg type relations. Here we present a quantum group generalization of bosonic string theory. Details are given in reference [5]. Associating a $SU(2)$-quantum group for each space-time dimension yields

$$[\alpha_m{}^A, \alpha_n{}^B] = \delta_{n+m}\eta^{AB}\frac{\cos\gamma n}{n}(\frac{\sin\gamma n}{\gamma})^2.$$

Here $\gamma$ is the deformation parameter (anisotropy parameter for statistical models). For $\gamma = 0$ we recover the usual Heisenberg commutations relations. $\eta^{AB}$ stands for the Minkowski space-time metric. Hence, we construct a one parameter continuous deformation of the usual string models. The scalar vertex operators for closed strings for an $SU(2)$ quantum group read $V(z,\bar{z},p) = V_L(z)V_R(\bar{z})$, where $z = \exp 2i(\sigma + \tau)$.

$$V_L(z) =: \exp ip \cdot X_L(z) : \quad V_R(\bar{z}) =: \exp ip \cdot X_R(\bar{z}) :,$$

$$X_L{}^A(z) = \frac{1}{2}q^A - \frac{i}{2}p^A \ln z + \frac{i\gamma\sqrt{\alpha'}}{2}\sum_{n\neq 0}\frac{z^{-n}\alpha_n{}^A}{\sin n\gamma}.$$

We interpret here $X^A(\sigma,\tau) = X_L{}^A(z) + X_R{}^A(\bar{z})$ as the string coordinate. It obeys the usual wave equation on the world sheet $(\partial_\sigma^2 - \partial_\tau^2)X(\sigma,\tau) = 0$ but the commutation relations depend on $\gamma$ even at equal times $(\tau)$. Therefore we find for the canonical conjugate momentum an expression non-local in the world-sheet. The equations of motion are derived from the action

$$S = \frac{1}{8\pi\alpha'}\int_0^\pi d\sigma d\tau \sum_{n=0}^\infty (-1)^n \partial_\mu X^A(\sigma_n^+,\tau)\partial^\mu X^A(\sigma_n^-,\tau),$$

which is non-local in the world-sheet. The $\gamma$-deformation has split the points in the kinetic energy by a distance $2\sigma_n = (n+1/2)$ with $z \in \mathbb{Z}$. This action is translational invariant on $(\sigma,\tau)$ but lacks manifest reparametrization invariance. The equations of motion are fully conformally invariant. The conserved tensor $T_{\mu\nu}$ is given by

$$T_{\pm\pm} = 4\sum_{n=0}^\infty (-1)^n \partial_\pm X^A(\sigma_n^+)\partial_\pm X^A(\sigma_n^-), \quad T_{+-} = T_{-+} = 0$$

where $X_+ = (\sigma + \tau)$. $T_{++}$ can be expanded as

$$T_{\pm\pm} = 8\alpha' \sum_{n\in\mathbb{Z}} \exp 2in(\tau+\sigma) \sum_l L_n^l,$$

where

$$L_n^l = \frac{(-1)^l}{2}\sum_{p\in\mathbb{Z}}\frac{p(l-p)\gamma^2}{\sin\gamma p \sin\gamma(l-p)}\cos[(2p-n)(l+1/2)\gamma]\alpha_p\alpha_{n-p}.$$

They satisfy $L_n^l = L_n^{-l-1}$. These $L_n^l$ are deformations of the usual Virasoro operators. When $\gamma \to 0$ we recover the usual $L_n$ and $T_{\mu\nu}$ operators [6]. The $L_n^l$ satisfy the following algebra up to a central term

$$[L_n^r, L_m^s] =$$
$$\frac{(n-m)}{4}(L_{n+m}^{s-r}\beta_{m,n}^{r,s+1} + L_{n+m}^{r-s}\beta_{n,m}^{s,r+1} + L_{n+m}^{r+s}\beta_{m,n}^{r,-s} + L_{n+m}^{r+s+1}\beta_{m,n}^{r+1,-s-1})$$
$$+ \frac{1}{4}(-M_{n+m}^{s-r}\rho_{m,n}^{r,s+1} + M_{n+m}^{r-s}\rho_{n,m}^{s,r+1} + M_{n+m}^{r+s}\rho_{m,n}^{r,-s} + M_{n+m}^{r+s+1}\rho_{m,n}^{r+1,-s-1}),$$

## 14. Quantum Group generalizations of String Theory

where $\beta_{mn}^{rs} = \cos[\gamma(mr+ns)]$, $\rho_{mn}^{rs} = \sin[\gamma(mr+ns)]$ and

$$M_n^l = \frac{(-1)^l}{2} \sum_{p \in \mathbb{Z}} \frac{p(n-p)\gamma^2 \alpha_p \alpha_{n-p}(2p-n)}{\sin\gamma p \sin\gamma(n-p)} \sin[\gamma(2p-n)(l+1/2)].$$

The effect of the deformation on the Virasoro algebra transforms it into a three-index infinite algebra. The $N$-scalar tree amplitude for the present quantum group generalized strings are

$$A_M(\mathbf{k}) = \int d\mu_M^\gamma(\mathbf{y}) l_M(\mathbf{k}, \mathbf{y}),$$

where $\mathbf{k} = (k_1, \ldots, k_M)$, and

$$I_{M\,\text{open}}(\mathbf{k}, \mathbf{y}) = \prod_{1 \leq i < j \leq M} (y_i^2 + y_j^2 - 2y_i y_j \cos\gamma)^{k_i \cdot k_j / 2},$$

$$l_{M\,\text{closed}}(\mathbf{k}, z\bar{z}) = \prod_{1 \leq i < j \leq M} |z_i^2 + z_j^2 - 2z_i z_j \cos\gamma|^{k_i \cdot k_j / 4}.$$

(The momenta $k$ are on shell $k_i^2 = -m^2$, $k_{\text{open}}^2 = 2$, $k_{\text{closed}}^2 = 8$). The two-point function involved here is

$$\langle X^A(y), X^B(y') \rangle = \frac{-1}{2} \eta^{AB} \log[1 + (\frac{y'}{y})^2 - 2\frac{y'}{y} \cos\gamma].$$

The quantum group deformation removes the usual equal point singularities. Singularities split in complex conjugate pairs

$$z = z' e^{+i\gamma}, \quad \text{that is,} \quad (\sigma - \tau) - (\sigma' - \tau') = +\gamma/2.$$

The deformed string amplitudes are not invariant under homographic transformations in $y_j(z_j)$ for $\cos\gamma \neq 1$. Only the dilatation invariance survives the quantum group deformations. The determination of the measure $d\mu_\gamma$ is given in [5]. We define the tree amplitude as $\tilde{A}_M(k) = A_M(\mathbf{k})/N_\delta$, $N_\delta$ being the volume of the dilatation group. We find

$$\tilde{A}_M(\mathbf{k})_{\text{open}} = $$
$$\int \ldots \int x_2^{M-2} x_3^{M-3} \cdots x_{M-1} dx_2 dx_3 \ldots dx_M \, I_M(\mathbf{k}, \mathbf{W}),$$

$$\tilde{A}_M(\mathbf{k})_{\text{closed}} = $$
$$\int \ldots \int [|z|_2^{M-2} |z|_3^{M-3} \ldots |z|_{M-1}]^2 d^2z \ldots d^2z_M \, I_M(\mathbf{k}, \mathbf{W}),$$

where $\mathbf{W} = (1, z_1, z_2 z_3, \ldots, z_2 \cdots z_M)$. Notice that here only one integration variable gets fixed ($z \equiv 1$) instead of three in the usual string amplitudes. Both closed and open amplitudes turn out to be symmetric functions of the external momenta: $A_M(\mathbf{k}) = A_M(\mathbf{k}_p)$ where $\mathbf{k}_p = (k_{p_1}, k_{p_2}, \ldots, k_{p_m})$, $p \in S_M$. The four point amplitude is

$$\tilde{A}_4(s,t)_{\text{open}} =$$

$$\int_0^1 \int_0^1 \int_0^1 dx\, dy\, dz\, y^{-s/2-1} [\lambda(x)\lambda(z)]^{-s/4-1},$$

$$\times [\lambda(xy)\lambda(yz)]^{1+(s+t)/4} [\lambda(y)\lambda(xyz)]^{-t/4-1},$$

$$\tilde{A}_4(s,t)_{\text{closed}} =$$

$$\int d^2x\, d^2y\, d^2z\, |y|^{-2-s/4} [\lambda(x)\lambda(z)]^{-2-s/8}$$

$$\times [\lambda(y)\lambda(xyz)]^{-2-t/8} [\lambda(xy)\lambda(yz)]^{2+(s+t)/8}.$$

Here $s = -(k^1 + k^2)$, $t = -(k^2 + k^3)^2$ as usual, and $\lambda(x) = |1 + x^2 - 2x\cos\gamma|$. These amplitudes are symmetric: $\tilde{A}_4(s,t) = \tilde{A}_4(t,s)$ and real for real $s, t$. Their unique singularities are simple poles at $s = 2n$, $n = 0, 1, 2, \ldots$ (open), or $s = 8n$, $n = 0, 1, 2, \ldots$ (closed). The residues are polynomial of degree $n$ [$2n$] in $t$ corresponding to states with spin $J \leq n$ [$2n$] for open [closed] strings. Here, the leading Regge trajectory is $s = 2J$ (open), $s = 4J$ (closed), $J = 0, 1, 2, \ldots$. We see that there are no poles at the tachyon mass. Moreover, the whole leading trajectory of the usual string [6] ($s = 2(J-1)$ or $s = 4(J-2)$) does not appear in the tree amplitude singularities. In other words, the $\gamma$-deformation shifts the intercept to $s = 0$. When $\gamma$ is purely imaginary, the $\lambda$ factors also contribute to the singularities; they transform part of the simple poles into double and triple poles.

The high energy behaviour of the four-point amplitudes ($s \to \infty$, $t$ fixed) is given by

$$\tilde{A}_4(s,t)_{\substack{\text{open } s\to\infty \\ t \text{ fixed}}} = \frac{1}{|s|}(4\sin^2\gamma/2)^{-1-t/4}$$

$$\times \int_0^1 \int_0^1 \frac{dx\, dz}{(1-xz)[1+xz-(x+z)\cos\gamma]} \left[\frac{\lambda(x)\lambda(z)}{\lambda(xz)}\right]^{1+t/4} + O(\frac{1}{s^2}).$$

These amplitudes do not exhibit Regge behaviour but a Born-type behaviour in this regime. The limits $s \to \infty$ and $\gamma \to 0$ do not commute. This reflects the modification of the short distance structure for

$\gamma \neq 0$. In spite of the presence of an infinite number of poles in Regge trajectories, the scattering amplitudes do not have Regge behaviour for $s \to \infty$. For $s \to \infty$, $t \to \infty$ with $s/t$ fixed, we find

$$\tilde{A}_4(s,t)_{\text{open } s,t \to \infty} = \frac{16(1-\cos\gamma)}{s^2 \cos\theta(1-\cos\theta)} \log\left[\frac{|s|(1-\cos\theta)}{8(1-\cos\gamma)}\right] + O(\frac{1}{s^3}),$$

where $\cos\theta = 1 + 2t/s$. Once more this behaviour is different from the very soft behaviour of the usual strings.

We have reported here a quantum group generalization of string amplitudes and studied their properties. These are the first steps to investigate the physical relevance of the models constructed in this note. This opens several direction of research.

## References

[1] See for reviews: P.P. Kulish and E. K. Sklyanin, in *Lecture Notes in Physics* **151**, Springer, 1981.

   H. J. de Vega, *PAR-LPTHE* preprint and references therein.

[2] V.G. Drinfel'd, *Dokl. Akad. Nauk* **32** (1985)25;

   M. Jimbo, *Lett. Math. Phys.* **11** (1986)247 and *Lett. Math. Phys.* **10** (1985)63.

[3] I. B. Frenkel and N. Jing, *Yale preprint*, 1988.

[4] V. G. Drinfel'd, *Dokl. Akad. Nauk* **296** (1987)135.

[5] H. J. de Vega and N. Sánchez, *preprints PAR-LPTHE* 88-30 and *Meudon-DEMIRM* 88-101.

[6] M. Green, J.H. Schwartz and E. Witten, *Superstring Theory*, vol. 1 and 2, Cambridge Univ. Press, 1987.

LPTHE, UNIVERSITÉ PIERRE ET MARIE CURIE, PARIS, FRANCE.
DEMIRM, OBSERVATOIRE DE PARIS, SECTION DE MEUDON, FRANCE.

# 15

## Towards a Covariant Closed String Theory

*J.G. Taylor and A. Restuccia*

### 1. Introduction

A very elegant construction of a covariant field theory for open strings and superstrings was given by Witten [1], based on the pioneering use by Siegel of the 1st quantised BRST charge $Q$ [2]. The extension of the string-field product to closed strings has been attempted by Strominger [3]. However the string product now has the defects of (a) being non-associative (b) generating a background dependent kinetic energy (c) being non-local in the embedding space-time (d) giving a multiple cover of moduli space in the resulting multi-loop amplitudes. Other attempts using $\Phi$ have difficulties with string length [4]. The approach proposed here will be based on the more geometrical attempt to build a general relativity of loop spaces. This has been developed by others [5] and ourselves [6]. In the earlier of these references difficulties were experienced in the introduction of a string field product. This is because the natural product in such an approach is $[\Phi(X)]^n$, where $n = 2, 3, \ldots$ and $\Phi$ is some functional of the string variables $X^\mu(\sigma)$, $0 \leq \sigma \leq 2\pi$, $\mu = 1, \ldots d$. This was avoided in [1] but then the discussion was restricted to open strings, so that the problems noted above arising in the approach of [3] did not occur. It is the purpose here to indicate briefly how the approach of [6] may provide a more satisfactory method of resolving some of the difficulties (a)-(d) above (only strings are considered, though the extension to superstrings appears straightforward).

### 2. Symmetries

For a $d$-dimensional smooth manifold $M$, the loop manifold, denoted $LM$, is the set of loops $X^\mu(\sigma)$ as above. Then the symmetry Diff $S^1 \times (GCT)$ (where $(GCT)$ is the pull-back to $LM$ of the general coordinate transformations $GCT$ on $M$) is a symmetry of the Nambu action

$$\int\int_W g^{\alpha\beta}\partial_\alpha X^\mu \partial_\beta X^\nu G_{\mu\nu}(X) d^2\sigma, \tag{1}$$

with $W$ being the two-dimensional world sheet with metric $g$, in a manifold $M$ with metric $G$, and $\alpha, \beta = 0, 1$. It is proposed to consider the larger symmetry $G = (\text{Diff } S^1)_{\text{loc}} \times (GCT)_{LM}$. The first factor corresponds to loop reparametrisations which depend on the loop,

$$\sigma \mapsto \bar{\sigma}(\sigma, X), \qquad (2)$$

whilst the second to co-ordinate transformations of the loop variable $X^M = \{X^\mu(\sigma)\}_{\sigma \in S^1}$ as

$$X^M \mapsto \bar{X}^N(X^M). \qquad (3)$$

This latter symmetry (3) is an extension of that above, $(GCT)$, and does not leave the Nambu action invariant. It may be that (3) is broken spontaneously down to (1), and become evident at high temperature [7]. Since it should allow a background-independent approach, choice of a background would also lead to breakdown of (3), as noted for a smaller symmetry group by Bardakci [5].

The associated tangent space group to $G$ will be

$$\mathfrak{g} = \text{KMISO}(d-1,1) \times \text{Vect } S^1 \qquad (4)$$

where KM denotes Kac-Moody; this has been noted independently in [8]. The corresponding multibein will be $E_M^A(x)$ and spin connections $W_M^{[ab]\sigma}, W_M^\sigma$, with algebra for the generators of $\mathfrak{g}$ given by

$$T(f, \Lambda, r) = \qquad (5)$$
$$\int d\sigma [\Lambda^{[ab]\sigma} X^{a\sigma} \delta/\delta X^{b\sigma} + f(\sigma) X^{a\sigma} \delta/\delta^{a\sigma} + r^{a\sigma} \delta/\delta X^{a\sigma}]$$

has commutator

$$[T(f_1, \Lambda, r_1) T(f_2, \Lambda, r_2)] = \qquad (6)$$
$$T([f_1, f_2], [\Lambda_1 \Lambda_2] + \Lambda'_{[1} f_{2]}, r_{[1} \Lambda_{2]} + r_{[1} f_{2]}),$$

where $[f_1, f_2] = f_1' f_2 - f_1 f_2'$ (the usual bracket in Vect $S^1$). This leads to the usual torsions and curvatures $R$, such as

$$R_{MN}^\sigma = \partial_{[M} W_{N]}^\sigma + W^\sigma{}_{[M} * W_{N]}^\sigma. \qquad (7)$$

## 3. The Product

It is necessary to define the string product entering in (7). This will be done by using a close analogy to that for the light cone gauge. This latter is known to require various contact terms [9], and so it is proposed to restrict the set of loops $X$ under consideration to those which only have at most a single (possible multiple) self-intersection. To see what such a restriction achieves, the properties of a product defined without it will be considered.

Let $I(X) = \{$set of self intersection points of $X$ in $LM\}$ and $|I(X)|$ be the number of points in $I(X)$ (counted according to multiplicity), then $LM$ decomposes naturally as $LM = \bigcup_{n \geq 0} L_n M$ where $L_n M = \{X : |I(X)| = n\}$. Any string field on $LM$ has a decomposition $\Phi = \Phi_n$, with $\Phi_n$ defined on $L_n M$.

It is now possible to introduce a natural product $*$ on the set of string fields $\Phi$ by

$$\Phi * \Psi(X_n) = \sum_r \Phi_r(Y_r) \Psi_{n-r-1}(Z_{n-r-1}), \qquad (8)$$

where the summation on the r.h.s. of (8) is over all choices $Y, Z$ such that $Y_r \cdot Z_{n-r-1} = X_n$ (the $\cdot$ product being the one used in homotopy). The resultant product is non-associative, since $(\Phi * \Psi) * \Xi$ defined on $X_0 \cdot Y_0 \cdot Z_0$ (where $X_0$ and $Y_0$ have a single intersection point, as do $Y_0$ and $Z_0$) does not contain $\Phi(Z_0)\Psi(X_0)\Xi(Y_0)$ on the r.h.s. of (8) if $X_0 \cap Y_0 = \emptyset$ whilst $\Phi * (\Psi * \Xi)$ does not contain such a term. However on restricting $X$ to only have a single self-intersection point with $|I(X)| > 1$ then it may be seen [6] that the $*$ product is now associative and commutative. For orientated strings (where account must be taken of a parametrisation split between different sub-loops of a given loop) the resulting product is still associative, but now non-commutative.

## 4. Constrained Geometry

To construct suitable constraints (to reduce the large class of fields present) and possible actions it appears useful to introduce an invariant metric on $S^1$. If $e^\phi$ is defined to have conformal weight 2 then $e^{\phi/2} d\sigma$ will be invariant. An extension of the manifold $M$, and the associated geometry, is now required to include $\phi$ (which is a bosonised ghost). It is possible to identify the usual string field as $E^{\phi\alpha}_{\phi\nu} = \Phi(x)\delta^\alpha_\nu$, and construct an action.

Somewhat natural constraints then appear [6]. However this final part of the programme is not yet satisfactory; the use of the condition of respresentation conservation, so helpful in supergravity and supersymmetry, seems required here to give direction to constraint choices.

## Acknowledgements

One of us (A.R.) would like to thank the British Council for financial support.

## References

[1] E. Witten, *Nucl.Phys.* **B268** (1986) 253; **B276** (1986)291.

[2] W. Siegel, *Phys. Lett.* **B151** (1983)391-396.

[3] A. Strominger, *Nucl. Phys.* **B294** (1987) 93.

[4] P.C. Bressloff, A. Restuccia, and J.G. Taylor, *Class. and Q. Grav.*, to appear.

[5] K. Bardakci, *Nucl. Phys.* **B271** (1986)561; **B284** (1987)334; **B297** (1988)583;
I. Bars and S. Yankielowicz, *Phys. Rev.* **D35** (1987)3878; *Phys Lett.* **B196** (1987)329;
M. Kaku, *Int. J. Mod. Phys.* **A2** (1987)1; *Phys. Lett.* **B200** (1988)22.

[6] A. Restuccia and J.G. Taylor, *Phys. Lett.* **B213** (1988)11.

[7] D. Gross, Superstrings, invited talk, *EPS Conference*, Munich, 1988.

[8] L. Castellani, *Phys. Lett.* **206** (1988)46.

[9] J.R. Greensite and F.R. Klinkhamer, *Nucl. Phys.* **B281** (1987)269; **B291** (1987)557; *LBL preprint* 23830 1987;
M.B. Green and N. Seiberg, *Nucl. Phys.* **B299** (1988)559;
A. Restuccia and J.G. Taylor, *Phys. Lett.* **B 213** (1988) 16; *Int. J. Mod, Phys.* **3** (1988)2855; *Phys. Rep.* **174** (1989)285.

DEPARTMENT OF MATHEMATICS, KING'S COLLEGE, STRAND, LONDON WC2R 2LS, ENGLAND.

DEPARTMENT OF PHYSICS, UNIVERSIDAD SIMON BOLIVAR, CARACAS 1080A, VENEZUELA.

# 16

## Contact Interactions for Light-cone Superstrings

A. Restuccia and J.G. Taylor

### 1. Introduction

It was pointed out recently [1] that the cubic interactions of the original light-cone gauge superstring field theory were incomplete. The addition of a quartic term $\hat{H}_4$ was required in order to ensure that the total Hamiltonian, now $(\hat{H}_2 + \hat{H}_3 + \hat{H}_4)$ instead of $(\hat{H}_2 + \hat{H}_3)$, be positive (where $\hat{H}_n$ denotes a contribution to the L.C. field theory Hamiltonian of degree $n$ in the fields). This latter is a condition guaranteed by [10]-SUSY; its violation would correspond to a violation of [10]-SUSY, which, however, the theory is supposed to possess. The construction of the full [10]-SUSY algebra was proposed in [1], but not completed there. These questions were analysed in [2] and more fully in [3]. The present situation has been summarized in [4]. It is the purpose of this article to survey these developments.

### 2. Closing the [10]-SUSY algebra

The linearly realised sub-algebra of the super-Poincaré algebra now includes the generators $Q_i^{+A}$, $Q_i^{+\bar{A}}$, for i= 1,2 equal to the zero modes of $\lambda^A, \bar{\lambda}^A, \theta^A, \bar{\theta}^A$ where $\lambda^A, \theta^{\bar{A}}$, are the R-moving 4 and $\bar{4}$ of SU(4) arising from an SO(8) 8-spinor using an $SU(4) \times U(1)$ decomposition of SO(8) as in [5]. The total angular momentum acquires extra contributions from the Grassmann-valued variables $\lambda, \theta, \bar{\lambda}, \bar{\theta}$. The non-linearly realised generators extend to the super-charges $\hat{Q}^{-A}$, $\hat{Q}^{-\bar{A}}$, which satisfy the [10]-SUSY algebra.

The introduction of an interaction $\hat{H}_3$ is expected only to be possible provided that there is an associated insertion factor $V_3^H$ (at the interaction point) and similar insertion factors $V_3^Q, V_3^Q$ for $\hat{Q}_3^{-A}, \hat{Q}_3^{-\bar{A}}$ (and $\hat{J}_3^{i-}, J_3^{+-}$ ). The insertion factors for the former operators were constructed by Green and Schwarz [5].

At cubic order it is necessary to check

$$[\hat{Q}_3^{-\bar{A}}, \hat{Q}_3^{-B}]_+ = [\hat{Q}_3^{-A}, \hat{Q}_3^{-B}]_+ = 0$$
$$[\hat{Q}_3^{-\bar{A}}, \hat{Q}_3^{-\bar{B}}]_+ = 0. \qquad (1)$$

The evaluation of the four external massless state matrix elements of (1) may be done directly to give [3, 4]

$$\langle 12|[\hat{Q}_3^{-\bar{A}}, \hat{Q}_3^{-B}]_+|34\rangle = \int \langle |\Psi^4 \cdots |\rangle \int_{C_{12}} d\tilde{\rho}_{12} \frac{\theta_1^{(\bar{A}} \theta_2^{B)}}{\sqrt{4F_1'' F_2''}}$$
$$\times \begin{cases} (4\bar{F}_1'' \bar{F}_2'')^{-1/2} \\ (4\bar{F}_1'' \bar{F}_2'')^{-5/2} \end{cases} \qquad (2)$$

for heterotic and type II respectively. $F$ is a map from the world sheet $\rho$ to the Riemann surface $z$; $\tilde{\rho}^i$ is the ith interaction point and $\tilde{\rho}_i = F_i$. The dots denote non-singular quantities, and the contour denoted $C_{12}$ for variable $\tilde{\rho}_{12} = \tilde{\rho}_1 - \tilde{\rho}_2$ is given in terms of the source $x$, (the other sources being at $0, 1, \infty$) by a contour in the $z$ plane of roughly two circles (one inside the other) and a radial line joining a point $A$ on one circle to $B$ on the other in the tree case (with similar features for higher loops). The points $A$ and $B$ correspond to the situations when the interaction points in the $Q_3$'s coincide. The contour $C_{12}$ is traversed in equal and opposite directions, so would apparently give zero. However the denominator functions $F_1'', F_2''$ (arising from the insertion factors), and from $(\det \partial_\rho)^{-12}$ (in the heterotic case) vanish when $\tilde{\rho}_{12} = 0$, so at the points $A, B$, since these values of $x$ correspond to double zeros of $F''$, not single zeros. The integrands of (2) are therefore poorly defined. Regularisation is necessary. The most general approach is to cancel all singular terms in (2). The singularity in (2) is 0 for the heterotic, $\tilde{\rho}_{12}^{-1}$ for type II. In the latter case it is therefore necessary to introduce a non-zero $\hat{Q}_4^{-A}$, so as to cancel (2) by the term $[\hat{Q}_2^{-\bar{A}}, \hat{Q}_4^{\bar{B}}]_+$. $Q_4^A$ has to be non-zero. It is also necessary to consider (1) with ([3, 4])

$$\langle 12|[\hat{Q}_3^{-\bar{A}}, \hat{Q}_3^{-B}]_+|34\rangle = \begin{cases} \delta^{\bar{A}B} \int (\frac{dx}{x} + \cdots) \\ \delta^{\bar{A}B} \int \frac{dx}{|x|_4 + x} + \text{non } \delta^{\bar{A}B} \text{ terms} \end{cases} \qquad (3)$$

in the heterotic and type II cases respectively. Equation (3) indicates again the need for $\hat{Q}_4^{-\bar{A}} \neq 0$ for the type II, whilst it allows $\hat{Q}_4^{-\bar{A}} = 0$ but requires $\hat{H}_4 \neq 0$ for the heterotic. Difficulties are encountered in (3) from terms which do not seem possible to be cancelled whatever regulation is used.

## 3. Conclusions

The construction of the fully [10]-SUSY algebra is still incomplete for the heterotic string due to the difficulties of dealing with the $J^{I-}$-components of the algebra. The extension of the mode techniques of [6] seem to lead to identities which are difficult to prove. However functional methods have been developed which should allow the full closure to be achieved [7]. The situation is not so clear-cut for the types I or II superstrings, and it may be for them that it is not possible to close the algebra at all. On the other hand alternative (covariant) techniques may be more appropriate in that case, although that would appear to require the development of a fully covariant superstring field theory to be fully justified.

## Acknowledgements

One of us (A.R.) would like to thank the British Council for financial support.

## References

[1] J. Greensite and F.R. Klinkhamer, *Nucl. Phys.* **B281** (1987)269; **B291** (1987)557; Superstring Amplitudes and Contact Interactions, *LBL preprint* 23830, August 1987.

[2] M. B. Green and N. Seiberg, *Nucl. Phys.* **B299** (1988)559.

[3] A. Restuccia and J.G. Taylor, *Phys. Lett.* **B213** (1988)16; *Int. J. Mod. Phys.* **A3** (1988) 2855.

[4] A. Restuccia and J.G. Taylor, Superstrings in the light-Cone Gauge", *Phys. Rept.* **174** (1989)285.

[5] M. B. Green and J.H. Schwarz, *Nucl. Phys.* **B243** (1984) 475.

[6] N. Linden, *Nucl. Phys.* **B286** (1987)429.

[7] A. Restuccia and J.G. Taylor, *in preparation.*

DEPARTMENT OF PHYSICS, UNIVERSIDAD SIMON BOLIVAR, CARACAS 1080A, VENEZUELA.
DEPARTMENT OF MATHEMATICS, KING'S COLLEGE, STRAND, LONDON WC2R 2LS, ENGLAND.

# 17

## Closed String Field Theory: the Failure of BRST Cohomology for the Open String

*P. Mansfield*

We are used [1] to identifying the physical configurations of the open string with the cohomology classes generated by the BRST charge, $Q$. We will argue that this fails as soon as we allow interactions, but that this failure can be exploited to construct a string field theory that describes closed strings [2].

We begin by describing the usual identification. All the dynamics of the free, open, bosonic string are contained in the Virasoro gauge conditions, which are restrictions on the physical states, together with the Virasoro algebra

$$(L_n - \delta_{n,0})|\,\text{physical}\rangle = 0, \quad n \geq 0,$$

$$[L_n, L_m] = (n-m)L_{n+m} + \frac{26}{12}(n^3 - n)\delta_{n,-m}.$$

The $L_n$ are constructed from the space-time coordinates describing the string. We can obtain a neater formulation by introducing anti-commuting degrees of freedom, the Faddeev-Popov ghost and antighost fields $c, b$, and defining the BRST charge

$$Q = \sum_n c_n(L_n - \delta_{n,0}) + \frac{1}{2}\sum_{nm}(n-m)\!:\!c_n c_m b_{-n-m}\!:.$$

The Virasoro algebra is equivalent to the nilpotency of $Q$, $Q^2 = 0$, and for every physical state we can find a state in the space enlarged to accommodate $c$ and $b$ such that

$$Q\,|\,\text{physical}\rangle = 0,$$

which we now use as the physical state condition. Because $Q$ is nilpotent we make another physical state by adding a state of the form $Q|\chi\rangle$.

$$|\,\text{physical}'\rangle = |\,\text{physical}\rangle + Q|\chi\rangle,$$

$$Q \mid \text{physical}'\rangle = 0.$$

$Q|\chi\rangle$ is called a spurious state because it decouples from physical matrix elements

$$\begin{aligned}\langle \text{ physical}''| \text{ physical}'\rangle &= \langle \text{ physical}''|\{ \mid \text{physical}\rangle + Q \mid \chi\rangle\} \\ &= \langle\text{physical}''| \text{physical}\rangle.\end{aligned}$$

Spurious states are interpreted as unphysical, gauge degrees of freedom. Thus, two states differing by a spurious state seem to describe the same physics and should be considered equivalent. So we are lead to identify the physical configurations of the string with the equivalence classes of the physical states, i.e.

$$\text{``Physics''} = Ker\ Q\ /\ Im\ Q.$$

Similarly, operators of the form $[Q, V]$ decouple from physical matrix elements, and are known as spurious operators.

$$\langle \text{ physical} \mid [Q,V] \mid \text{physical}' \rangle = 0.$$

The identification of physics with the cohomology classes of $Q$ is no longer appropriate as soon as we allow open strings to interact. Open strings interact locally by the world-sheet of a single string splitting into two. The interacting theory is elegantly described by Witten's string field theory action [3] constructed from the string field A

$$S_W = \int (AQA + A^3).$$

The first term gives the free equation of motion $QA = 0$, the second describes the three string interaction. The $n$-string scattering amplitude is given (in the usual LSZ formalism for this action) by the vacuum expectation value

$$\mathcal{A}_n = \langle \prod_{p=1}^{n} \int \Psi_p QA \rangle_{S_W},$$

where the $\Psi_p$ are the wavefunctions of the scattered open strings, so $Q\Psi_p = 0$. The interaction leads to processes in which open strings propagate in loops. The simplest is the one loop diagram. In this case the world-sheet of the virtual string consists of an annulus. The world-sheets of the scattered open strings are then attached to sections of the boundaries of the annulus. Witten's theory reproduces the usual first-quantised expression for this amplitude. This is an integral [4] over

17. Closed String Field Theory                                         177

the single real modulus, $\tau$, that parametrises non-trivial deformations of the complex structure on the annulus. The external states are then represented by vertex operators placed on the boundary of the annulus. $\tau$ can be taken to be the ratio of the radii. In this case the boundary of moduli space consists of the two configurations $\tau = 0$, which is a disc punctured at the centre, and $\tau = 1$, which is an annulus of infinitesimal thickness.

Now this one loop process may be interpreted differently. It is also the transition amplitude for a closed string to propagate from a state specified by the boundary conditions (and vertex operator insertions) on the inner boundary of the annulus to a state specified by those on the outer one. Since closed strings propagate in open string loops, any theory of open strings necessarily contains closed strings. This was the way closed strings were originally encountered in the dual resonance model [5]. Within the sigma-model approach we can compute mixed scattering processes that involve both open and closed strings simply by placing closed string vertex operators on the body of the open string world-sheet. This leads to the question: What are the physical open string states $\Psi$, with $Q\Psi = 0$, that describe closed strings in that they can be used in the LSZ formalism for Witten's theory to compute these mixed amplitudes? We cannot identify them with the cohomology classes of $Q$ because these are exhausted by the usual open string states. Neither can we simply enlarge the open string space of states to include the usual closed string space of states as this would be double counting. It would be equivalent to adding a closed string field into $S_W$, and the point is that the theory already contains closed string interactions. We must look for these states amongst the physical open string states we already have. We can find them, but we must give up the identification of physical configurations with the cohomology classes of the BRST charge $Q$.

There is another problem associated with interactions which turns out to be crucial to obtaining open string states describing closed strings. This is the failure of operators of the form $[Q, V]$ to decouple from physical loop processes [6]. For the free theory we had

$$\langle \text{ physical}| [Q, V] |\text{physical}' \rangle = 0.$$

For a loop process the amplitude is the integral over the moduli of the Riemann surface of the sigma-model expectation value of a product of BRST invariant vertex operators

$$\mathcal{A} = \int_M \langle V_1 .. [Q, V] .. V_n \rangle_{\sigma-model}.$$

This time the expectation value does not vanish, so we cannot show that the complete expression is zero. All we can show is that the integrand is an exact form on moduli space, so

$$\mathcal{A} = \int_M d\langle\, V_1 .. \, \tilde{V} .. \, V_n\, \rangle = \int_{\partial M} \langle\, V_1 .. \, \tilde{V} .. \, V_n\, \rangle,$$

for some $\tilde{V}$. In general this will not vanish. In fact the boundary of moduli space contains Riemann surfaces where boundaries collapse to punctures, so we are left with contributions in which the vertex operator $\tilde{V}$ sits on a puncture. For the simple case of one open string loop, the integral over $\tau$ reduces to a sum of two terms corresponding to the boundary components $\tau = 1$ and $\tau = 0$. The first vanishes but the second is just what we need to describe a scattering process involving open strings and a closed string with a vertex operator $\tilde{V}$. We conclude that we can represent closed string vertex operators by certain spurious vertices $[Q, V]$.

Open string states can be obtained by computing the sigma-model functional integral for a semicircular world-sheet with the vertex operator inserted at the middle of the curved boundary. On this boundary we impose conditions on the dynamical degrees of freedom corresponding to free propagation. On the diameter, the space-time coordinates are held at fixed values equal to the argument of the wave-function. If we use a spurious vertex we obtain a spurious state. In fact the open string state describing a closed string tachyon of momentum $2k^\mu$ is given by the functional integral with the following insertions:

(1) $c^r b_{rs} \zeta^s$ at the middle of the curved boundary. $\zeta$ is an appropriately chosen vector field.

(2) Two insertions of open-string vertices, $e^{ik.X}$, integrated along the curved boundary.

(3) The BRST charge, which may be represented as the integral of the BRST current along the diameter.

The two $e^{ik.X}$ insertions are bona fide open string tachyons which together carry the closed string tachyon momentum $2k^\mu$. $\zeta$ is an appropriately chosen vector field. Using this in Witten's theory correctly generates scattering amplitudes involving closed string tachyons [2].

Since closed strings are described by certain spurious open string states we cannot consider all the elements of a cohomology class of $Q$ as being physically equivalent. The spurious states are not all just gauge degrees of freedom.

Having obtained these states it is now a simple matter to construct a string field theory that describes closed strings only. The usual topological expansion for a closed string theory only involves Riemann surfaces without boundaries. Since the boundaries where we attach our spurious states all collapse to punctures, we must simply find a way of truncating the theory so as to exclude diagrams with boundaries to which no spurious states are attached. This can be done by including fermionic degrees of freedom on the boundaries. These will contribute to any amplitude additional factors given by the functional integral

$$\int \mathcal{D}(\tilde{\psi},\psi) exp\,(i \oint dy^a \tilde{\psi} \partial_a \psi).$$

The point is that this action has zero-modes, $\tilde{\psi}$=constant, $\psi$=constant, so the integral will vanish unless we make an additional insertion of $\tilde{\psi}\psi$. If we only make this insertion in each spurious state then we achieve our goal of truncation. The resulting theory has the same non-Abelian gauge invariance as Witten's, and should inherit from that theory the good property that the Feynman diagram expansion provides a cell decomposition of moduli space.

### References

[1] M. Kato and K. Ogawa, *Nucl.Phys.* **B212** (1983) 443.

[2] P. Mansfield, *Nucl.Phys.* **B317** (1989) 187.

[3] E. Witten, *Nucl.Phys.* **B268** (1986) 79.

[4] M.B. Green, J.H. Schwarz and E. Witten, *Superstring Theory*, Cambridge University Press, Cambridge, 1987.

[5] G. Frye and L. Susskind, *Phys.Lett.* **31B** (1970) 589.

[6] P. Mansfield, *Nucl.Phys.* **B283** (1987) 551; E. Martinec, *Nucl.Phys.* **B281** (1987) 157.

DEPARTMENT OF THEORETICAL PHYSICS, 1 KEBLE ROAD, OXFORD, ENGLAND.

# 18

# Twistors and Four-dimensional Conformal Field Theory

*M.A. Singer*

This is a report (with technical details omitted) on work concerned with generalizations to four dimensions of two-dimensional CFT. Accounts of this and related material are contained in [5,9] and the articles of Hodges and Penrose in this volume. I thank these, my collaborators, and also Claude LeBrun and Graeme Segal for useful discussions.

To motivate the four-dimensional generalization, let us begin by recalling Segal's formulation [8] of 2-dimensional CFT. One considers pairs $(\Sigma, \iota)$, where

(a) $\Sigma$ is a compact Riemann surface with boundary $b\Sigma$ (and possibly choice of spin structure);

(b) $\iota$ is an identification of the boundary with a number of copies of the circle.

As a complex manifold, $\Sigma$, and so too its boundary, has an orientation. If the orientations of a boundary circle induced from $\Sigma$ and from $\iota$ agree, then we call the circle in-going; otherwise it is called outgoing. We assume tacitly that if $\Sigma$ is equipped with a spin-structure then $\iota$ also identifies the spin-structure of $b\Sigma$ with the standard spin-structure on the circle.

According to Segal's definition, a 2-dimensional CFT is a 'natural rule' which assigns an amplitude

$$A_{(\Sigma,\iota)} : \mathcal{H}^{\otimes p} \otimes \bar{\mathcal{H}}^{\otimes q} \longrightarrow \mathbf{C} \tag{1}$$

to each $(\Sigma, \iota)$ with $p$ in-going and $q$ out-going boundary circles, where $\mathcal{H}$ is a Hilbert space independent of $(\Sigma, \iota)$.

The very least that is meant by 'natural' is that $A_{(\Sigma,\iota)}$ should depend only upon $(\Sigma, \iota)$ up to the obvious notion of 'parametrized conformal equivalence' under which $(\Sigma, \iota)$ and $(\Sigma', \iota')$ are equivalent if $\Sigma$ and $\Sigma'$ are conformally equivalent and the equivalence relates $\iota$ and $\iota'$ in the obvious way. Other 'natural' properties relate to the

the behaviour of the $A$'s under orientation reversal and under the natural contraction operations of attaching an in-going to an outgoing boundary circle.

The basic example of a two-dimensional CFT arises as follows. Let $H = H_+ \oplus H_-$ be the Hilbert space of $L^2$ spinors on the circle, polarized in the standard way. Write $H^{p,q}$ for the direct sum of $p+q$ copies of $H$, polarized so that $H_\pm^{p,q}$ is the direct sum of $p$ copies of $H_\pm$ and $q$ copies of $H_\mp$. If $W_\Sigma$ is the space of $L^2$ spinors on $\Sigma$ which are holomorphic in the interior $\Sigma - b\Sigma$, $\iota$ identifies $W_\Sigma$ with a closed linear subspace $W = W_{(\Sigma,\iota)}$ of $H^{p,q}$.

To explain how this defines an amplitude up to scale, suppose for the moment that $\dim H < \infty$. Then the space of holomorphic sections of $\mathrm{Det}^*$ on $\mathrm{Gr}(H^{p,q})$ is $\mathcal{H}^{\otimes(p+q)}$ where $\mathcal{H}$ is the dual of the exterior algebra of $H$; and if $W$ is a linear subspace of $H^{p,q}$ then evaluation of sections at $W$ defines a map

$$\mathcal{H}^{\otimes(p+q)} \longrightarrow \mathbf{C}$$

up to scale. (The scale arises from the unnatural isomorphism $\mathrm{Det}^*_W \simeq \mathbf{C}$.)

In the infinite-dimensional setting, the situation is broadly analogous, subject to some important technical conditions [7,10]. First $\mathrm{Gr}(H^{p,q})$ is not the set of all closed subspaces $W \subset H^{p,q}$; one must confine oneself to those subspaces 'sufficiently close to' $H_+^{p,q}$. That is, $\mathrm{Gr}(H^{p,q})$ consists of those closed subspaces $W$ for which

the orthogonal projection $W \longrightarrow H_+^{p,q}$ is Fredholm,
and
the orthogonal projection $W \longrightarrow H_-^{p,q}$ is trace-class.

Secondly, although one can define a holomorphic determinant line bundle $\mathrm{Det}$ on the infinite dimensional complex manifold $\mathrm{Gr}(H^{p,q})$, the space of holomorphic sections of $\mathrm{Det}^*$ is not a Hilbert space—instead, it contains $\mathcal{H}^{\otimes p} \otimes \bar{\mathcal{H}}^{\otimes q}$ as a dense subspace, where $\mathcal{H}$ is the completion of the exterior algebra on $H_+ \oplus \bar{H}_-$. This is sufficient, however, for evaluation at $W$ to define an amplitude up to scale as in (1). A proof that $W$ actually is a point in $\mathrm{Gr}(H^{p,q})$ can be found in [7, p. 159]; the key tool is the expansion of elements of $W$ in Laurent series on the boundary circles.

There is obvious scope for generalization of this construction. What we have in mind is the replacement of the circle by some other fixed compact manifold $S$ and the replacement of the Riemann surfaces by some other class of manifolds $M$ with boundary equal to a union of

## 18. Twistors and Four-dimensional CFT

copies of $S$. In addition to these geometric objects, it will be necessary to find appropriate 'classical fields' on $M$, as generalizations of the holomorphic functions on $\Sigma$.

From the point of view of Riemannian differential geometry, one of the natural generalizations of a conformal structure on a surface is a conformally flat structure on an $n$-manifold. (Recall that such a structure is defined by an atlas whose changes of coordinates are conformal motions of $S^n$.) So let us consider pairs $(M, \iota)$ such that

($\alpha$) $M$ is a compact, oriented, conformally flat four-manifold with boundary (and possibly spin-structure);

($\beta$) $\iota$ is an identification (conformal isometry) of $bM$ with a number of copies of $S^3$ with its standard conformal structure.

Once again, we distinguish between in-going and out-going copies of $S^3$ and implicitly assume that $\iota$ identifies spin-structures as appropriate. A four-dimensional CFT is then defined to be a 'natural rule' which assigns an amplitude

$$A_{(M,\iota)} : \mathcal{H}^{\otimes p} \otimes \bar{\mathcal{H}}^{\otimes q} \longrightarrow \mathbf{C} \qquad (2)$$

to each $(M, \iota)$ with $p$ in-going copies and $q$ out-going copies of $S^3$. Once again, $\mathcal{H}$ is supposed to be a Hilbert space independent of $M$.

To give an example of such a rule, we shall apply the above Grassmannian construction with the classical fields being solutions on $M$ of the spin-$s$ conformally invariant Dirac equation. (The spin-0 'Dirac equation' is the conformal Laplace equation.)

So let $H$ be the Hilbert space of $L^2$ initial data on $S^3$ for the Dirac equation (of some arbitrary but fixed spin $s$), polarized in the obvious way. With $H^{p,q}$ defined as before, each $(M, \iota)$ with $p$ in-going and $q$ out-going boundary components determines a subspace $W = W_{(M,\iota)}$ of boundary values of $L^2$ solutions of the Dirac equation on $M$. As before, if $W$ is a point of $\mathrm{Gr}(H^{p,q})$, then it determines an amplitude up to scale. There are two things to check: that the map $\iota^*$, which assigns to a solution of the Dirac equation its boundary values, is injective; and that $W$ is 'close' (in the usual sense) to $H_+^{p,q}$.

These matters can adequately be treated using elliptic theory for the Dirac operator, but we wish to sketch how they can be discussed by using complex analysis. This is achieved by means of twistor theory [1,2,6].

Recall that any conformally flat four-manifold $M$ has a twistor space $Z$: this is a complex 3-manifold which fibres over $M$, the fibres being holomorphically embedded Riemann spheres. The basic example is the map $\pi : \mathbf{C}P^3 \to S^4$ which is best understood by thinking of the four-sphere as the quaternionic projective line. Since projective motions of $\mathbb{C}P^3$ which preserve $\pi$ are in one-one correspondence with conformal motions of $S^4$, the existence of $Z$ for any $M$ now follows from our earlier definition of conformally flat structure. (In fact, the complex manifold $Z$ can also be constructed if $M$ is self-dual [1]; it would be interesting to try to develop a four-dimensional CFT for this larger class of four-manifolds.) Note that the set in $Z$ that corresponds to a 'round' three-sphere in $M$ is a real hypersurface $P$: when $M$ is $S^4$, so $P \subset \mathbb{C}P^3$, one can choose homogeneous coordinates so that $P$ is given by the vanishing of

$$|z_0|^2 + |z_1|^2 - |z_2|^2 - |z_3|^2.$$

Thus $P$ is not a complex manifold but a $CR$-manifold (cf. [4, Ch. V] where such manifolds are called 'partially complex').

Under this twistor correspondence, existence and choice of spin structure on $M$ correspond precisely to existence and choice of a fourth root of the line bundle $K$ of holomorphic 3-forms on $Z$; and $W_M$ corresponds to the analytic cohomology group $H^1(Z, \mathcal{O}(K^{(s+1)/2}))$ with appropriate boundary conditions [3]. With these facts at hand, one can prove the required properties of $W_{(M,\iota)}$: the Laurent expansion is here replaced by an expansion in 'elementary states' [3], these being cohomology classes which are singular on one projective line.

As before, such a rule is natural in that it is a 'parametrized conformal invariant' of $(M, \iota)$: it is an interesting problem to try to calculate it explicitly. To do so one would need an existence theory for cohomology classes on $Z$ with prescribed poles on lines, analogous to the standard existence theory for meromorphic functions on a Riemann surface. The development of such a theory involves many interesting questions of conformal and complex geometry.

We have discussed an apparently very natural generalization of CFT from two to four dimensions. It remains to comment briefly on certain aspects of the construction. First, we should note that the Hilbert space $\mathcal{H}$ of the four-dimensional theory has a natural interpretation in terms of massless spinor fields on real Minkowski space. This is very obvious from the twistor point of view, since the boundary $CR$-manifold $P$ is precisely the space of light rays in real compactified Minkowski space $\mathbf{M}$ and it is a result of [2] that $CR$-cohomology classes on $P$ correspond to massless fields on $\mathbf{M}$. It follows that with just this slight shift of viewpoint, all the amplitudes can be regarded as defined on Hilbert spaces built from Lorentzian spinor fields. Thus the twistor picture provides a kind of halfway house between the Lorentzian and Euclidean field theories. The Lorentzian point of view is more strongly emphasized in [5].

Our final comments are far more speculative. In two-dimensional CFT, the freedom in parametrization of a given boundary circle is a copy of the group $\text{Diff}^+(S^1)$ of orientation-preserving diffeomorphisms of the circle. This was allowable because if two Riemann surfaces are attached at a common boundary circle by any such diffeomorphism, the resulting surface still has a complex structure. Closely related is the fact that certain unitary representations of loop groups automatically define two-dimensional conformal field theories [8].

In dimension $n > 2$ the conformal group is finite-dimensional and this is reflected by the finite-dimensional freedom in the parametrizations of $S^3$ that were allowed in our definition of four-dimensional CFT. It is tempting to hope, nonetheless, that an appropriate generalization of the ideas presented here might supply a framework for the study of the corresponding infinite-dimensional Lie groups $\text{Diff}^+(S^3)$ and $\text{Map}(S^3, G)$. Such a goal is, however, very distant at present.

## References

[1] M.F. Atiyah, N.J. Hitchin and I.M. Singer, Self-duality in four-dimensional Riemannian geometry, *Proc. Roy. Soc. London* **A362** (1978) 425–461.

[2] M.G. Eastwood, R. Penrose and R.O. Wells Jr., Cohomology and massless fields, *Commun. Math. Phys.* **78** (1981) 305–351.

[3] M.G. Eastwood and A.M. Pilato, The density of twistor elementary states, *Pacific J. Math.*, 1989, to appear.

[4] G.B. Folland and J.J. Kohn, The Neumann problem for the Cauchy-Riemann complex, *Ann. of Math. Studies* **75**, P.U.P., Princeton 1972.

[5] A.P. Hodges, R. Penrose and M.A. Singer, A twistor conformal field theory for four space-time dimensions, *Phys. Lett.* **B216** (1989) 48–52.

[6] R. Penrose and W. Rindler, *Spinors and space-time*, (two volumes), C.U.P., Cambridge, 1984, 1986.

[7] A. Pressley, and G.B. Segal, *Loop groups*, Oxford Mathematical Monographs, O.U.P., Oxford, 1986.

[8] G.B. Segal, The definition of conformal field theory, 1989, *to appear*.

[9] M.A. Singer, Flat twistor spaces, conformally flat manifolds and conformal field theories in four dimensions, *preprint*, Oxford, 1989.

[10] E. Witten, Quantum field theory, grassmannians and algebraic curves, *Commun. Math. Phys.* **113** (1988) 529–600.

MATHEMATICAL INSTITUTE, 24-29 ST. GILES', OXFORD OX1 3LB, ENGLAND.

# 19

## Projective and Superconformal Structures on Surfaces

*W.J. Harvey*

Much attention has recently been given to the study of *super-Riemann surfaces*, by which we mean Riemann surfaces $S$ provided with an additional structure tantamount to a bundle of exterior algebras $E \cong \mathbf{C} \oplus \wedge^1 L$ with $L$ a holomorphic line bundle over $S$ such that $L \otimes L$ is isomorphic to the canonical cotangent bundle $K$ of $S$. Detailed accounts of these objects and their infinitesimal deformation theory can be found in [1], [2], [6], [8], [9] where they are fitted into the framework of complex supermanifolds, superconformal structures and graded sheaves.

One difficulty, which seems even more of a barrier than in the case of classical deformations of Riemann surface structure, is the lack of a good *global* description of super-moduli spaces. In this note, we outline an approach which places the theory in the classical setting of projective structures on variable Riemann surfaces. We explain how to construct a distribution (family of vector subspaces) inside the holomorphic cotangent space to the moduli space $\mathcal{M}_g(4)$ of Riemann surfaces with genus $g$ and furnished with a level-4 homology structure, such that the corresponding rank-$(2g-2)$ complex vector bundle models the soul deformations of a family of super-Riemann surfaces $(S_t, E_t)$ $t \in \mathcal{M}_g(4)$ whose underlying spin bundles $L_t$ are odd and non-special, that is, $H^0(S_t, \mathcal{O}(L_t)) \cong \mathbf{C}$ for all $t$. The keystone in this construction is the existence of holomorphic sections for the space of non-singular odd theta characteristics on $\mathcal{C}_g(4)$, the universal curve over $\mathcal{M}_g(4)$.

### 1. Projective structures on surfaces

**1.1** $\mathbf{P}^1$ denotes the complex projective line and $G = PGL(2, \mathbf{C})$ the conformal group of holomorphic automorphisms of $\mathbf{P}^1$. A *projective structure* on a (compact) surface $S$ is a collection of co-ordinate charts $\phi_\alpha : U_\alpha \to \mathbf{P}^1$, with $\{U_\alpha\}$ covering $S$, such that the transition mappings $f_{\alpha\beta}$ defined for overlapping $U_\alpha, U_\beta$ by $f_{\alpha\beta}(\phi_\alpha(p)) = \phi_\beta(p)$ lie in $G$. In terms of the universal covering $\tilde{S}$, this data pieces together to give a mapping $f: \tilde{S} \to \mathbf{P}^1$ known as the *developing map* of the projective structure; there is furthermore a *monodromy homomorphism* $\rho: \pi_1(S) \to G$, from the fundamental group of $S$ operating as deck transformations of $\tilde{S}$, such that $f(\gamma(x)) = \rho_\gamma(f(x))$ for each $\gamma \in \pi_1(S)$ and $x \in \tilde{S}$.

**1.2** We concentrate on surfaces of higher genus, $g \geq 2$, and we need to consider varying surfaces and projective structures on them. A detailed study may be found in [3] or [7].

It is well known that the source of projective structures is the space $Q = Q(S)$ of quadratic differentials on $S$: if $q \in Q$, then one solves in $S$ the differential equation $\{f, z\} = q(z)$, where the left side denotes the Schwarzian derivative $(f''/f')' - \frac{1}{2}(f''/f')^2$. This produces a holomorphic local homeomorphism $f$; conversely $\{f, z\}$ will always be a quadratic differential on $S$.

The Bers embedding theorem provides a holomorphic family $F_g \xrightarrow{\pi} T_g$ of Jordan domains $\mathcal{D}_t = \pi^{-1}(t)$, with base the Teichmüller space $T_g$ and discrete subgroups $\Gamma_t \subset G$ operating on $\mathcal{D}_t$ discontinuously to produce surfaces $S_t = \mathcal{D}_t/\Gamma_t$; this is a holomorphic family of (relative) projective structures on the family $V_g$ of Riemann surfaces $\{S_t, t \in T_g\}$.

The holomorphic cotangent bundle to $T_g$ comprises the totality $Q = \bigcup_t Q(S_t)$ of quadratic differential forms on the varying surfaces $S_t$. If we apply the procedure above to $q \in Q$ with suitable normalization – solutions to the Schwarzian equation are unique up to composition with elements of $G$ – then there results a local bijection from $Q$, viewed as parametrising the family of all projective structures $f$ on surfaces $S$ up to projective equivalence, into the space of monodromy homomorphisms $\rho = \rho(f) : \pi_1(S) \to G$ modulo conjugation of $\rho$ by elements of $G$. By a theorem due to Hejhal, rendered into more explicit form in [3], [7], this rule is a complex analytic local homeomorphism $F: Q \to \mathrm{Hom}(\pi_1(S), G)/G$, whose derivative at $\rho$ can be described in terms of (Eichler) cohomology groups. The tangent space to $\mathrm{Hom}(\pi_1(S), G)/G$ at $\rho$ is $H^1(\Gamma, \mathcal{P}_2)$, with $\rho(\pi_1(S)) = \Gamma \subset G$ acting on the space $\mathcal{P}_2$ of quadratic polynomials by the rule, for $\gamma \in \pi_1(S)$, $P \in \mathcal{P}_2$, $P \cdot \gamma = (P \circ \rho_\gamma) \cdot (\rho'_\gamma)^{-1}$. Earle derives an explicit splitting of the tangent space

$$0 \to Q(\Gamma) \to H^1(\Gamma, \mathcal{P}_2) \to Q(\Gamma)^* \to 0 \tag{1}$$

with kernel representing the vertical tangent vectors (tangent to the fibre of $Q \to T_g$) and image the tangent space to $T_g$.

## 2. Spin structures on surfaces.

**2.1** One can regard a spin structure on $S$ as an isomorphism class $[L]$ of holomorphic line bundles $L \to S$ such that $L \otimes L \cong K$, the

## 19. Projective and Superconformal Structures

canonical cotangent bundle of $S$. For a family of surfaces, in particular for the universal family $V_g \xrightarrow{\pi} T_g$, a spin structure is a (class of) holomorphic line bundle on $V_g$, which restricts on each fibre surface $S_t$ to a spin bundle $L_t$. There are $2^{2g}$ distinct spin structures on each surface $S$, corresponding to points of order 2 in the Jacobi variety of $S$ or alternatively to elements of $H^1(S, \mathbf{Z}/2)$; the set $\Sigma_g$ of spin structures on $V_g$ has the same description, *via* a suitable process of holomorphic continuation. According to the parity value of the quadratic form induced on spin structures by cup product on cohomology of $S$, the structures are labelled *even* or *odd*; a spin structure $[L]$ has the same parity as the dimension of $H^0(S, \mathcal{O}(L))$, its space of global holomorphic sections.

**2.2** For general families of surfaces, one needs to bring into play the modular group $Mod_g \cong \mathit{Diff}(S)/\mathit{Diff}_0(S)$, which changes the topological marking of the base surface. There is a permutation action of $Mod_g$ on the set $\Sigma_g$ which preserves parity and is transitive on the subsets of odd and even ones. A subgroup $Mod_g{}^{spin}$ of finite index in $Mod_g$ fixes all spin structures; it is in fact the inverse image, under the natural homomorphism onto the symplectic modular group $Sp(2g, \mathbf{Z})$, of the congruence subgroup of level 2 [11].

**2.3** The modular group operates discontinuously on the spaces $T_g$ and $V_g$, as biholomorphic automorphisms; the quotient is the *modular family* $\mathcal{C}_g \to \mathcal{M}_g$ which carries a natural complex analytic $V$-manifold structure. In the appropriate extended sense, one can consider the relative canonical $V$-bundle of this family and its set of (square root) spin bundles. The *spin-modular* family $\mathcal{C}_g{}^{spin} \to \mathcal{M}_g{}^{spin}$ is the ramified covering corresponding to the subgroup $Mod_g{}^{spin}$; thus over almost every point $t \in \mathcal{M}_g$ represented by a Riemann surface $S_t \subset \mathcal{C}_g$ there are finitely many holomorphically equivalent surfaces $S_t^j$, corresponding to labellings of the various spin structures on $S_t$. These families form the basic *body spaces* out of which the super-modular families of super-Riemann surfaces are built.

**2.4** Complex line bundles, such as spin-bundles $L \to S$ are conveniently described by systems of *multiplier functions* on the universal covering $\tilde{S}$; these are collections $\{e_\gamma, \gamma \in \pi_1(S)\}$ of non-vanishing holomorphic functions on $\tilde{S}$, satisfying the condition

$$e_{\gamma_1 \gamma_2}(z) = e_{\gamma_1}(\gamma_2(z)) \cdot e_{\gamma_2}(z).$$

For spin bundles, the multipliers satisfy $(e_\gamma(z))^2 = 1/\gamma'(z)$. The bundle $B$ determined by the multipliers $\{e_\gamma\}$ is defined by identifying points of $\tilde{S} \times \mathbf{C}$ under the group action of $\pi_1(S)$ given by the rule $\tilde{\gamma}(z, u) = (\gamma(z), e_\gamma(z)u)$.

**2.5** Meromorphic sections of $B$ always exist: their divisor class in Pic($S$) is denoted by $[B]$. A natural question to ask about a spin bundle over a space such as $T_g$ or $\mathcal{M}_g$ is whether there is a global *holomorphic* section; this would be a valuable aid in further study. In general (for $g \geq 5$) the answer is negative [4]. However a combination of the Lefschetz embedding theorem and Riemann's singularity analysis of the theta function implies that there are sections over Zariski-open sets of $\mathcal{M}_g$; see for instance [5], [10]. This leads to a more precise result which we only state here.

**Theorem** *There is a covering of the modular curve $\mathcal{C}_g(4)$ by finitely many open sets $V_\alpha = \pi^{-1}(U_\alpha)$ with spin bundles $L^\alpha$ such that, over each $t \in U_\alpha$, the restriction $L_t^\alpha$ of the (odd) spin bundle $L^\alpha \to \mathcal{C}_g(4)$ has a holomorphic section $s_t^\alpha$, $s_t^\alpha : S_t \to L_t^\alpha$ depending holomorphically on $t$. On intersections $U_\alpha \cap U_\beta$, the sections either coincide or are distinct everywhere.*

We remark that this provides a holomorphically varying family of half canonical positive divisors, a valuable resource for study of meromorphic forms and functions *via* the prime form.

## 3. Deformations and super-structures.

**3.1** The Kodaira-Spencer theory of infinitesimal deformations may also be applied to line bundles over surfaces; briefly, one deforms a bundle $B$ as described in **2.4** by specifying a 1-*cocycle* $\{\psi_\gamma\}$ of the group $\pi_1(S)$ with values in $\mathcal{O}(B)$, which is a collection of functions $\psi_\gamma(z)$ for $z \in \tilde{S}$, $\gamma \in \pi_1(S)$, satisfying

$$\psi_{\gamma\delta}(z) = \psi_\gamma(\delta(z)(e_\gamma(z))^{-1} + \psi_\delta(z), \text{ for all } \gamma, \delta \in \pi_1(S).$$

The deformation $B^\psi$ is the bundle quotient of $\tilde{S} \times \mathbf{C}$ by the group action

$$\tilde{\gamma}^\psi(z, u) = (\gamma(z), e_\gamma(z)(u - \psi_\gamma(z))).$$

Now we apply this construction to the contravariant spin bundle $L^{-1}$, whose multiplier system is $\sqrt{\gamma'(z)}$ (for some choice of square roots). *Via* Serre duality or the construction of potential functions for appropriately defined Beltrami forms, one can identify the space of 1-cocycles for $L^{-1}$ with the global holomorphic forms of weight $\frac{3}{2}$ – more precisely, with the space $H^0(S, \mathcal{O}(K \otimes L))$. Because $L \otimes L \cong K$, this latter space is isomorphic to $Q[L]$, the space of quadratic differential forms on $S$ with divisor a multiple of $[L]$. In view of the theorem in **2.5** this extends to the modular family over $\mathcal{C}_g(4)$, determining a holomorphic family $Q(\mathcal{L})$ of $(2g-2)$-dimensional subspaces of the cotangent bundle $Q_g(4)$ of $\mathcal{M}_g(4)$.

**3.2** The theory of projective structures in **1.2** can now be applied to the family of Riemann surfaces over $\mathcal{M}_g(4)$ furnished with holomorphic sections of the odd spin bundles. The projective structures on each $S_t$, $t \in \mathcal{M}_g(4)$, extend naturally to super-conformal structures by defining a differential operator $D_\tau$ for $\tau \in Q[L_t]$ by: $D_\tau = \partial/\partial\theta_\tau + \theta_\tau\, \partial/\partial z_\tau$ on sections of the bundle $\mathcal{E}$ of exterior algebras $E_t = \wedge L_t$; here $(z_\tau, \theta_\tau)$ are local co-ordinates for the bundles of projective lines over $S_t$ determined by the projective structures. Alternatively, one can pull back the canonical graded Lie algebra structure on $\mathbf{CP}^1$ to obtain graded sheaves on the families $Q(\mathcal{L})$, an approach developed in [12].

**3.3** The local deformation structure of the family described here fits inside the sequence (1) of **1.2** to give an infinitesimal splitting at each body surface $S = \mathcal{D}/\Gamma$,

$$0 \to Q(\Gamma) \to T(Q(\mathcal{L}))_\Gamma \to Q[L]^* \to 0. \qquad (2)$$

This indicates an intricate melding of real degree 1 and degree 2 Eichler cohomology for $\Gamma$ within the complex degree 2 cohomology of the quasi-fuchsian group $\Gamma$.

## References

[1] M.A. Baronov, I.V. Frolov, Yu.I. Manin and A.S. Schwarz, *Comm. Math. Phys.* **111** (1987)373-392.

[2] L. Crane and J. Rabin, *Comm. Math. Phys.* **113**(1988)610.

[3] C.J. Earle, *Annals of Math. Studies* **97**(1981)87-99

[4] C.J. Earle and I. Kra, *J. Math. Kyoto Univ.* **26-1** (1986)39-64.

[5] G. Gonzalez-Diez, *King's College Ph.D. thesis*, 1987.

[6] L. Hodgkin, *Lett Math Phys* **14** (1987)47-53.

[7] J. Hubbard, *Annals of Math. Studies* **97** (1981)257-275.

[8] C. Le Brun and M. Rothstein, *Comm. Math. Phys.* **117** (1988)159.

[9] E. Martinec, *Nucl. Phys.* **B281** (1987)157.

[10] D. Mumford, *Tata Lectures on Theta* Vol II, Birkhauser, Boston 1984.

[11] P. Sipe, *Math. Ann.* **260** (1982)67-92.

[12] P. Teofilatto, *PA natural sheaf of graded Lie algebras over variable Riemann surfaces*, King's College preprint.

DEPARTMENT OF MATHEMATICS, KING'S COLLEGE, STRAND, LONDON WC2R 2LS, ENGLAND.

# 20

# Unified Spin Gauge Theories of the Four Fundamental Forces

*J.S.R.Chisholm and R.S.Farwell*

## 1. Principles of Spin Gauge Theories

*Spin gauge theories* are Lagrangian field theories describing fundamental fermions and their interactions. Their properties have been developed over a period of years [1]–[10]. The Lagrangian density is defined in terms of elements of a Clifford algebra $C_{p,q}$, where $p + q = n$. In the models we are discussing, the base space of the Clifford algebra is the tangent space to an n-dimensional manifold which consists of two parts: (i) A four-dimensional curved space-time submanifold, with coordinate system $x \equiv \{x^\mu : \mu = 1, 2, 3, 4\}$ on a patch; (ii) An $(n-4)$-dimensional flat 'higher space'. The tangent space $T(x)$, $x \in M$ is spanned by a vector basis $\{e_i\}$ of $C_{p,q}$, with

$$\{e_i, e_j\} = 2I g_{ij}, \tag{1}$$

where $I$ is the unit scalar of $C_{p,q}$ and $(g_{ij}) = diag(p+, q-)$. An $x$-dependent vector basis on $M$ is given by

$$\Gamma_\mu(x) = h^i_\mu(x) e_i, \tag{2}$$

where $h^i_\mu(x)$ is the *vierbein field*. Then

$$\{\Gamma_\mu(x), \Gamma_\nu(x)\} = 2I g_{\mu\nu}(x) \tag{3}$$

with

$$g_{\mu\nu}(x) = g_{ij} h^{\ j}_\mu(x). \tag{4}$$

*Spinors* $\psi(x)$, representing fermions, are taken to be elements of minimal left ideals of the Clifford algebra. We shall use particular examples later. The *bar conjugate spinor* is

$$\overline{\psi}(x) = \psi^\dagger(x) \Gamma, \tag{5}$$

where the bar-conjugation matrix $\Gamma$ is defined to satisfy $e^\dagger \Gamma e \Gamma^{-1}$ and $\Gamma^\dagger = \Gamma$ for any vector $e$ of $C_{p,q}$. This ensures that $\overline{\psi} e \psi$ is real Hermitian.

Under spin gauge transformations, spinors and their conjugates have transformations of the form

$$\psi(x) \mapsto Q(x)\psi(x), \qquad \bar{\psi} \mapsto \bar{\psi}Q^{-1}(x) \qquad (6)$$

which implies that the conjugation matrix transforms by

$$\Gamma \mapsto [Q^\dagger(x)]^{-1}\Gamma(x)Q(x). \qquad (7)$$

The density matrix transformation is thus

$$\psi(x)\bar{\psi}(x) \mapsto Q(x)\psi(x)\bar{\psi}(x)Q^{-1}(x); \qquad (8)$$

since $\psi(x)\bar{\psi}(x)$ is not an ideal, but a more general element of the Clifford algebra, consistency within the algebra requires that a general element $A(x)$ transforms by

$$A(x) \mapsto Q(x)A(x)Q^{-1}(x). \qquad (9)$$

This transformation of an 'operator' is a fundamental difference between 'spin gauge theories' and standard gauge theories; it implies, for instance, that the 'constant' Dirac matrices $\{\gamma_\mu\}$ become $x$-dependent under local gauge transformations.

## 2. Tetrahedral Structure of Idempotents of the Algebra $C_{3,1}$

The algebra $C_{3,1}$ has four primitive idempotents (interpreted as spin-energy projection operators when the algebra is used as the space-time algebra). These idempotents exhibit the symmetry of a regular tetrahedron; this property generalises the 'triangle symmetry' of three idempotents, established by Greider and Weiderman [11].

The basis vectors $c_r; r = 1, 2, 3, 4$ of $C_{3,1}$ satisfy the anticommutation relations

$$\{c_r, c_s\} = 2g_{rs}I, \qquad (10)$$

where $g_{rs} = \text{diag}(+1, +1, +1, -1)$, and $I$ is now the unit of the algebra $C_{3,1}$. We choose the (real) primitive idempotents to be

$$\begin{aligned} P_1 &= 1/4(1 + c_3 c_4)(1 + c_1) \\ P_2 &= 1/4(1 + c_3 c_4)(1 - c_1) \\ P_3 &= 1/4(1 - c_3 c_4)(1 - c_1) \\ P_4 &= 1/4(1 - c_3 c_4)(1 + c_1) \end{aligned} \qquad (11)$$

satisfying

$$P_i P_j = \delta_{ij} P_i \qquad (i, j = 1, 2, 3, 4). \qquad (12)$$

## 20. Unified Spin Gauge Theories

The simplest $4 \times 4$ matrix representation of the idempotents is

$$P_i = (e_{ii}) \qquad (i = 1, 2, 3, 4), \tag{13}$$

where $(e_{ij})$ is a square matrix whose only non-zero entry is unity in the $i, j$ position. This representation is realised if we choose the basis vector representation

$$c_1 = \nu_3 \tau_3, \ c_2 = \nu_3 \tau_2, \ c_3 = \nu_2 \tau_4, \ c_4 = i\nu_1 \tau_4, \tag{14}$$

where $\{\nu_p\}$ and $\{\tau_p\}$ are two sets of $2 \times 2$ Pauli matrices, $\nu_4$ and $\tau_4$ are the corresponding unit matrices; (14) consists of direct products of matrices, the $\tau$-matrices being inserted as blocks into the $\nu$-matrices.

In order to introduce interchange relations, we use the notation $c_r c_s = c_{rs}$, $(r \neq s)$, and write $P_{r,s}$ to indicate the alternative '$P_r$ or $P_s$'. These relations are

$$\begin{aligned} c_{23} P_{2,3} c_{23}^{-1} &= P_{3,2}, & c_{23} P_{1,4} c_{23}^{-1} &= P_{4,1} \\ c_{31} P_{3,1} c_{31}^{-1} &= P_{1,3}, & c_{31} P_{2,4} c_{31}^{-1} &= P_{4,2} \\ c_{12} P_{1,2} c_{12}^{-1} &= P_{2,1}, & c_{12} P_{3,4} c_{12}^{-1} &= P_{4,3}. \end{aligned} \tag{15}$$

The three relations to the left were introduced (in a different representation) by Greider and Weiderman, who dealt only with the properties of $P_1$, $P_2$, $P_3$, which they used to represent quarks of three colours. We make a similar interpretation, but we extend their set of relations by including $P_4$; this idempotent is taken to represent the electron-neutrino or 'lepton'. If we picture $P_i$ ($i = 1, 2, 3, 4$) as non-adjacent vertices of a cube, as in Fig.1, we can interpret the interchange operation geometrically as rotations through $\pi$ about axes parallel to the centre of the cube, as shown. The four idempotents are then represented by the vertices of a regular tetrahedron; we shall show that the idempotents exhibit the symmetry properties of the tetrahedral group. We modify and generalize the 'colour permutation operator' of Greider and Weiderman by defining the invertible operator

$$B_4 = P_4 - c_{12} P_1 - c_{23} P_2 - c_{31} P_3. \tag{16}$$

We also differ from Greider and Weiderman by considering similarity transformations rather than left multiplication. Using (12) and (15), it is not difficult to show that

$$B_4 P_1 B_4^{-1} = P_2, \ B_4 P_2 B_4^{-1} = P_3, \ B_4 P_3 B_4^{-1} = P_1, \tag{17}$$

and

$$B_4 P_4 B_4^{-1} = P_4. \tag{18}$$

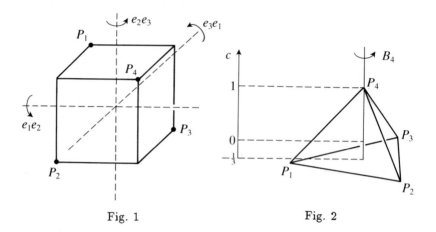

Fig. 1    Fig. 2

As a result of (2.), the operator $B_4$ can therefore be interpreted as the 'colour permutation operator' of the three quarks, represented geometrically by a rotation of the tetrahedron through angle $2\pi/3$ about an axis through $P_4$, as shown in Fig.2. The invariance (18) of $B_4$ accords with the tetrahedral picture, and with the interpretation of $P_4$ as the lepton, invariant under colour transformations.

Corresponding to rotations through $2\pi/3$ about axes $P_1$, $P_2$ and $P_3$, there are three operators similar to $B_4$:

$$\begin{aligned} B_1 &= P_1 - c_{12}P_4 - c_{23}P_3 - c_{31}P_2 \\ B_2 &= P_2 - c_{12}P_3 - c_{23}P_4 - c_{31}P_1 \\ B_3 &= P_3 - c_{12}P_2 - c_{23}P_1 - c_{31}P_4. \end{aligned}$$

It is not difficult to show that a similarity transformation by $B_1$ permutes $(P_2, P_3, P_4)$, and similarly for $B_2$ and $B_3$, so that all the tetrahedral symmetry properties have been identified. Further, we can establish the relations

$$B_4 B_1 B_4^{-1} = B_2, \quad B_4 B_2 B_4^{-1} = B_3, \quad B_4 B_3 B_4^{-1} = B_1,$$

and three similar sets, showing that the operators $\{B_i\}$ form a representation of the tetrahedral group. They also satisfy $B_i^3 = I$. In the representation (14), the colour permutation matrix $B_4$ takes the form

$$B_4 = \begin{pmatrix} 0 & 0 & -i & 0 \\ i & 0 & 0 & 0 \\ 0 & 1 & 0 & 0 \\ 0 & 0 & 0 & 1 \end{pmatrix}. \tag{19}$$

20. Unified Spin Gauge Theories

Naturally, this matrix has a block structure, with a unit $1 \times 1$ lepton block and a $3 \times 3$ quark block which permutes the colours. This $3 \times 3$ block is, as Greider and Weiderman pointed out, an SU(3) matrix element; they omitted the fourth row and column in their representation. The full set of Gell-Mann matrices are formed from the $4 \times 4$ matrices

$$\begin{aligned}
\Lambda_3 &= P_1 - P_2  & \Lambda_8 &= (P_1 + P_2 - 2P_3)/\sqrt{3} \\
\Lambda_1 &= i(P_1 B_4^2 - P_2 B_4) & \Lambda_2 &= P_1 B_4^2 + P_2 B_4 \\
\Lambda_4 &= i(P_1 B_4 - P_3 B_4^2) & \Lambda_5 &= P_1 B_4 + P_3 B_4^2 \\
\Lambda_6 &= P_3 B_4 + P_2 B_4^2 & \Lambda_7 &= i(P_3 B_4 - P_2 B_4^2)
\end{aligned} \quad (20)$$

by omitting the fourth rows and columns, which consist only of zeros.

## 3. Spin Gauge Theory Models

We shall illustrate the ideas that we have used by describing three models; the third model incorporates electroweak, gravitational, and SU(3) 'strong' interactions. We shall also show how a new concept of fermion mass, based on the frame field, gives rise to correct boson masses for photons, W and Z bosons, and gluons, without introducing the Higgs-Kibble mechanism.

### 3.1 The Frame Field Concept

We exemplify this concept by considering the Dirac equation

$$\gamma^\mu(i\partial_\mu - eA_\mu)\psi + m\psi = 0 \qquad (21)$$

In any representation, and on any curved space-time manifold, the upper- and lower-suffix Dirac matrices satisfy

$$\gamma^\mu(x)\gamma_\mu(x) \equiv 4U, \qquad (22)$$

where $U$ is the unit of the algebra $C_{1,3}$. So we can write (21) as

$$\gamma^\mu(x)[i\partial_\mu - eA_\mu)U + 1/4m\gamma_\mu(x)]\psi(x) = 0. \qquad (23)$$

The expression in square brackets is the *'extended covariant derivative'*; we note the analogy between i) the coupling constant $e$ and the potential $A_\mu(x)$ and ii) the constant $1/4m$ and the field $\gamma_\mu(x)$ which is $x$-dependent in a general gauge. We take this analogy quite seriously; since $\{\gamma_\mu(x)\}$ represents the space–time reference frame at each point $x$, we call it the *'Frame Field'*. We regard the frame field as a physical field; the mass constant $1/4m$ is the coupling of the fermion field to the frame field. In our view, mass is not an intrinsic property of a particle, but a kind of 'friction' between the particle and the frame field, making it travel at a speed less than its 'natural speed', the speed of light.

## 3.2 Interactions of the Electron-Neutrino

This model of electroweak and gravitational interactions of the electron-neutrino is based upon a 7-dimensional manifold $M$ consisting of two submanifolds: $M_1$ is 4-dimensional curved space-time, and $M_2$ is a flat 3–dimensional 'higher space'. We define the basis vectors of the tangent space $T(x)$ in terms of a set of Dirac matrices $\{\gamma_i; i = 1, 2, 3, 4\}$ with $\gamma_1^2 = \gamma_2^2 = \gamma_3^2 = -\gamma_4^2 = -U$ and $\eta = \gamma_1\gamma_2\gamma_3\gamma_4$ and a 'Pauli set' $\{\rho_r; r = 1, 2, 3, 4\}$; $U$ is the unit $4 \times 4$ matrix. The vector basis $\{e_i; i = 1, 2, \ldots 7\}$ of the algebra $C_{1,6}$ is taken to be the set of $8 \times 8$ matrices

$$e_i = \rho_4 \gamma_i \quad (i = 1, 2, 3, 4),$$
$$e_{i+4} = \rho_i \eta \quad (i = 1, 2, 3).$$

The frame field on $M$ is the set $\{\Gamma_\mu(x); \mu = 1, 2, 3, 4\}$, where

$$\Gamma_\mu(x) = \rho_4 \gamma_\mu(x) \equiv \rho_4 h_\mu{}^i \gamma_i. \qquad (24)$$

The lepton spinor is defined as the left ideal (an $8 \times 8$ matrix)

$$l = \begin{pmatrix} \epsilon & \underline{0} & \underline{0} & \underline{0} & \underline{0} & \underline{0} & \underline{0} & \underline{0} \\ \nu & \underline{0} & \underline{0} & \underline{0} & \underline{0} & \underline{0} & \underline{0} & \underline{0} \end{pmatrix}$$

where $\epsilon$, $\nu$ are the electron and neutrino 4-component bispinors, and $\underline{0}$ is the zero bispinor. If $\bar{\epsilon}$, $\bar{\nu}$ are the Dirac bar–conjugate bispinors, the conjugate lepton bispinor $\bar{l}$ is the right ideal ($8 \times 8$ matrix) containing the non-zero first row $(\bar{\epsilon}, \bar{\nu})$ and all other rows zero. We can write the fermion mass terms in terms of the frame field as in Section (3.1). In our next model, we shall explain the reason for writing the mass terms in this form.

The generators of the electroweak and gravitational gauge groups $SU(2)$, $U(1)$ and $SL(2, \mathbf{C})$ are respectively $\{\rho_a U\}$, $\rho_4 U$ and $\rho_4 \gamma_i \gamma_j$, $(i < j)$ with $a = 1, 2, 3$ and $i, j = 1, 2, 3, 4$. The three sets of generators commute with each other. The helicity projection operators are $h_\pm = (U \pm i\eta)/2$, and we take the group of spin gauge transformations to be

$$Q(x) = \exp\{i[g\rho_a h_+ \theta^a(x) + g'(\rho_3 h_- + \rho_4 U)\theta^4(x)] - i\rho_4 \gamma_i \gamma_j \theta^{ij}(x)\}.$$

In the free fermion Lagrangian we introduce the electroweak and gravitational spin connections

$$\Omega_\mu = [g\rho_a h_+ W_\mu{}^a(x) + g'(\rho_3 h_- + \rho_4 U) W_\mu{}^a(x)]/2, \qquad (25)$$

$$G_\mu = \rho_4 \gamma_i \gamma_j G_\mu{}^{ij}(x). \quad (i < j) \qquad (26)$$

### 20. Unified Spin Gauge Theories

If we introduce the Weinberg angle $\tan^{-1}(g'/g)$, and define the fields $A_\mu$ and $Z_\mu$ in terms of $W_{3\mu}$ and $W_{4\mu}$ in the usual way, the term

$$\mathrm{Tr}\{\bar{l}\Gamma^\mu \Omega_\mu l\} \tag{27}$$

in the Lagrangian density gives the standard electroweak interactions of the electron and its neutrino.

### 3.3 The Four Fundamental Interactions

In this section we discuss the first family of fermions. We combine the algebras $C_{1,6}$ of model in Section (3.2) and the algebra $C_{3,1}$ of Section 2 into the algebra $C_{4,7}$ of $32 \times 32$ matrices; these matrices are formed by replacing each element of the $4 \times 4$ matrices of $C_{3,1}$ by the product of that element with an $8 \times 8$ matrix of $C_{1,6}$. We write these direct products as simple products of $\{P_i, B_r, \nu_p \tau_q\}$ and $\{\rho_s \gamma_i, \ldots\}$.

The $8 \times 8$ lepton spinor is given in Section 3.2, and the quark spinors of three colours ($c = r, b, y \equiv 1, 2, 3$) are similarly defined by replacing $(\epsilon, \nu)$ by $(d_c, u_c)$. The $32 \times 32$ 'family spinor' is then represented by

$$\psi = \sum_c P_c q_c + P_4 l = \begin{pmatrix} q_1 & 0 & 0 & 0 \\ 0 & q_2 & 0 & 0 \\ 0 & 0 & q_3 & 0 \\ 0 & 0 & 0 & l \end{pmatrix}.$$

We define the basis vectors of $C_{4,7}$ by combining the bases $\{c_r\}$ in (14) and $\{e_i\}$ in Section 3.2; anti–commutation is ensured by introduction of the factor $\nu_3 \tau_1$ into 3.2. The full basis of $C_{4,7}$ is then

$$\begin{aligned} e_i &= \nu_3 \tau_1 \rho_4 \gamma_i & (i = 1, 2, 3, 4) \\ e_{i+4} &= \nu_3 \tau_1 \rho_i \eta & (i = 1, 2, 3) \\ e_{i+7} &= c_i \rho_4 U & (i = 1, 2, 3, 4). \end{aligned} \tag{28}$$

The close relationship of the last four vectors to $\{c_i\}$ ensures that the tetrahedral structure of Section 2 is unchanged, applying to $8 \times 8$ blocks within the $32 \times 32$ matrices. Also, the introduction of the factors $\nu_3 \tau_1$ into the first four vectors only introduces a unit factor $I = \nu_4 \tau_4$ into the terms in $Q(x)$, which are all in the even part of the algebra. We note that the term in the lepton interaction generated by $g' \rho_4 U \theta^4(x)$ in $Q(x)$ must acquire a factor $-1/3$ to give the correct quark interaction; to achieve this we introduce the 'particle charge operator'

$$C = P_4 - (P_1 + P_2 + P_3)/3. \tag{29}$$

Apart from the helicity factors, the generators of the gauge transformations are

| Electroweak | | Gravitational | Strong |
|---|---|---|---|
| SU(2) | U(1) | SL(2,C) | SU(3) |
| $I\rho_a U$ | $C\rho_4 U$ | $I\rho_4 \gamma_i \gamma_j$ | $\Lambda_l \rho_4 U$ |
| $a = 1,2,3$ | | $i < j$ | $l = 1,\ldots,8$ |

The full gauge group is

$$Q(x) = \exp\{-1/2i[gI\rho_a h_+ \theta^a(x) + g'(I\rho_3 h_- + C\rho_4 U)\theta^4(x)] \\ - iI\rho_4 \gamma_i \gamma_j \theta^{ij}(x) - if\Lambda_l \rho_4 U \Phi^l\}; \qquad (30)$$

invariance under this group is ensured if the derivative $\partial_\mu$ in the free fermion kinetic Lagrangian density is replaced by

$$D_\mu = \partial_\mu - \Omega_\mu - G_\mu - S_\mu, \qquad (31)$$

where

$$\Omega_\mu = i[gI\rho_a h_+ W_\mu{}^a(x) + g'(I\rho_3 h_- + C\rho_4 U)W_\mu{}^4(x)]/2,$$
$$G_\mu = iI\rho_4 \gamma_i \gamma_j G_\mu{}^{ij}(x),$$
$$S_\mu = if\Lambda_l \rho_4 U V^l,$$

with sums over $a = 1,2,3$; $i,j = 1,2,3,4$ $(i < j)$; $l = 1,2,\ldots,8$. The five terms in the exponential in (30) commute with each other, since $C$ is colour symmetric, and $P_4$ annihilates $\{\Lambda_l\}$.

### 3.4 The Frame Field and the Extended Covariant Derivative

We can introduce lepton and quark masses into the present model as in Section 3.1 by defining the 'extended covariant derivative'

$$\triangle_\mu = D_\mu + E_\mu, \qquad (32)$$

where

$$E_\mu = i\{(P_1 + P_2 + P_3)(\mu_1 \rho_3 + \mu_2 \rho_4)\gamma_\mu + P_4(\mu_3 \rho_3 + \mu_4 \rho_4)\gamma_\mu\}. \qquad (33)$$

Then the masses of the down and up quarks and the electron and its neutrino can be identified as

$$m_d = \mu_2 + \mu_1 \qquad m_u = \mu_2 - \mu_1$$
$$m_e = \mu_4 + \mu_3 \qquad m_\nu = \mu_4 - \mu_3.$$

The standard method of obtaining gauge-invariant contributions to the boson free Lagrangians is to form the self-commutator of $D_\mu$, to contract this self–commutator with itself, and to take the trace. We

## 20. Unified Spin Gauge Theories

adopt this method, but we replace the covariant derivative $D_\mu$ by the extended covariant derivative $\triangle_\mu$, forming

$$\operatorname{Tr}\{g^{\mu\rho}g^{\nu\sigma}[\triangle_\mu,\triangle_\nu][\triangle_\rho,\triangle_\sigma]\}. \tag{34}$$

Several different gauge-invariant Lagrangian density terms arise from (34); these can be normalised separately. We enumerate these using an abbreviated notation, multiplied by normalisation constants $N_1, N_2, N_3, N_4$:

1. $N_1|[D_\mu, D_\nu]|^2$: this term gives the free photon, W and Z boson, and gluon kinetic terms. It also gives the 'spin gravity' term proportional to $R^{\mu\nu\rho\sigma}$ quadratic in curvature.

2. $N_2|[D_\mu, E_\nu] + [E_\mu, D_\nu]|^2$ : this term gives boson mass terms; the photon and gluon masses are zero, and the $m_W/m_Z$ mass ratio is correct.

3. $N_3|[E_\mu, E_\nu]|^2$: this term is constant contribution to the Lagrangian density, and so is a cosmological term.

4. $N_4|[D_\mu, D_\nu][E_\mu, E_\nu]|$ : with a suitable choice of $N_4$, this is the Einstein–Hilbert gravitational Lagrangian density $R/16\pi G$.

In a complete theory, we would like the normalisation constants $\{N_p\}$ to be equal. This is not possible in this model, since $N_4$ depends upon $G$, and must be many orders of magnitude greater than $N_1$ and $N_2$. The frame field plays a central role in our theories. We enumerate the various concepts with which it is associated:

1. It generates the Dirac algebra, and is thus related to the spin and energy sign of fermions.

2. At each point of the space-time manifold, it represents the frame of reference.

3. By the basic Clifford algebra relation (3), it is the 'Dirac square root' of the metric tensor, and we call $\Gamma_\mu(x)dx^\mu$ the 'linear metric'. The space-time metric involves gravitation.

4. The frame field is introduced by factorising the fermion mass terms, leading to the interpretation of mass as the coupling constant of a fermion to the frame field.

5. The inclusion of the frame field in the extended covariant derivative provides mass terms for the W and Z bosons, in the correct ratio; this inclusion gives zero photon and gluon masses.

6. The extended covariant derivative also provides the Einstein-Hilbert gravitational Lagrangian, in addition to the 'spin gravity' term, and it provides a cosmological constant term; the spin gravity term gives a short-range modification of Einstein gravitation.

## Acknowledgements

We are pleased to acknowledge the influence of Greider and Weideman's paper, and thank them for sending us a preliminary copy. One of us (J.S.R.C.) benefitted from a period of work at the University of Adelaide, and from numerous discussions with Prof. C.A. Hurst and his group. We also thank colleagues in the University of Kent Applied Mathematics group for their continuing interest.

## References

[1] L. Halpern, *Proc. 1st Marcel Grosssman Meeting*, ed. R. Ruffini, North-Holland, 1977, p.113.

[2] J.S.R. Chisholm and R.S. Farwell, *Lecture Notes in Physics* **116** (1980), p.305.

[3] J.S.R. Chisholm and R.S. Farwell, *Proc. Roy. Soc. London* **A337** (1981)1.

[4] Z. Dongpei, *Phys.Rev.* **D22** (1980)2027.

[5] W.J. Wilson, *Phys.Lett.* **A75** (1980)172.

[6] J.S.R. Chisholm and R.S. Farwell, *Nuovo Cimento* **82A** (1984) 2; 145; 185; 210.

[7] A.O. Barut and J. McEwan, *Phys. Lett.* **B135** (1984)172; **B139** (1984) 464; *Lett. Math. Phys.* **11** (1986) 67.

[8] J.S.R. Chisholm and R.S. Farwell, *J. Phys.* **A20** (1987)6561.

[9] J.S.R. Chisholm and R.S. Farwell, *Gen. Rel. Grav.* **20**(1988)371.

[10] J.S.R. Chisholm and R.S. Farwell, *J. Phys.* **A22** (1989) 1059.

[11] K.Greider and T.Weideman, *UKC Seminar*, June 1987, and *UCD Preprint*, April 1988.

INSTITUTE OF MATHEMATICS, UNIVERSITY OF KENT, CANTERBURY, KENT, ENGLAND.

ST. MARY'S COLLEGE, TWICKENHAM, MIDDLESEX, ENGLAND.

# 21

## Path Integral Formulation of Chiral Gauge Theories

*T.D. Kieu*

One of the fundamental, and fascinating, problems of quantum field theories is that of anomalies. Whenever a classical-level symmetry is broken at the quantum level, an anomaly has occurred. These anomalous breakings of global symmetries entail no fatal consequence for the internal consistency of a theory. It is an entirely different situation, however, if the broken symmetry is a local gauge symmetry associated with either large or small transformations.

I will restrict myself in this talk to the case of chiral gauge theories, where the left-handed and the right-handed parts of the fermion current couple differently to the gauge fields. General reviews of anomalies can be found elsewhere in the literature [1].

As the gauge symmetry is the fundamental principle in the building of gauge field theories, it is undesirable to have the principle forfeited upon quantisation. Furthermore, this conflict results in some mathematical inconsistency of anomalous gauge theories [2,3,4]. The argument to be put forward here is that this should not be the case. There should be no anomaly associated with the gauge group; and there should be no inconsistency [5].

It suffices to quantize the theory of chiral fermions in a complex representation of the gauge group, abelian or non-abelian, with the gauge fields treated as classical. The path integral representation for the vacuum-to-vacuum amplitude is a functional of the gauge fields $A_\mu$ up to some (infinite) multiplicative constant. There is an ambiguity in the action of the path integral $\mathcal{Z}[A]$: the action is the classical action modulo some term dependent on $A_\mu$. This term, being $\psi$-independent, cannot change the Feynman rules involving fermion lines. The evaluation of the famous triangle diagram thus exhibits, as usual, the anomalous nature of the divergence equation. Even so, the theory is consistent, as it should be, until the gauge fields are quantized. Then, the consistency condition that $\mathcal{Z}[A]$ is invariant under a gauge transformation, either large or small, of the gauge fields has to be imposed. The condition fixes the ambiguous term in the action, up to some gauge invariant term, to be the Wess-Zumino (WZ) term [6], defined by its variation under a gauge transformation to be the anomaly, but with an appropriate sign to cancel the contribution from the fermion part. Thus, the fermion current is only part of the total 'current' coupled to the gauge fields —albeit non-local counterterms have appeared.

It is shown below how the WZ term is realised in a second-quantisation approach to derive the path integral.

In the temporal gauge, I expand the spinor wavefunctions in terms of the instantaneous normalizable eigenstates of the first-quantized *interacting* hamiltonian. Usual properties of orthonormality and completeness for the expanding basis are assumed. This hamiltonian varies with time via the time dependence of the gauge fields. I also impose the periodic condition on the hamiltonian for the time interval $[0, T]$, $\mathcal{H}(0) = \mathcal{H}(T)$, where $T$ is to be taken to infinity later. Such time-dependent periodic gauge fields correspond to a closed loop in the infinite-dimensional manifold of all static gauge-field configurations with $A_0 = 0$. To apply the adiabatic approximation, I have further assumed that there is no degeneracy (level crossing) along this closed loop. Note that the single-valuedness of the expanding basis $\{\phi_E\}$ is not presupposed. Such a preconditioning is not necessary, in fact, and can be wrong, as it turns out.

As in the usual procedure of second quantisation, the coefficients of the above expansion become creation/annihilation operators with their canonical equal-time anti-commutators. The Fock space is built by iterative use of these operators on the vacuum, which is defined to have all the negative-energy levels filled. The second-quantized hamiltonian is normal ordered with respect to this vacuum.

To obtain the generating functional, I employ the approximation of the time-ordered evolution operator from 0 to $T$ by the product of $N$ operators of infinitesimal time interval $\Delta t$, $N \Delta t = T$. The path integral is readily derived by the insertion of instantaneous completeness expressions for the coherent states [7]. Alternatively—and equivalently—I use the holomorphic representation [8] for such purpose. The representation is on holomorphic functions of Grassmann variables where the inner product is that of Berezin.

The result, from either way, resembles the text-book-standard path integral of free field theory since I have diagonalized the fully interacting hamiltonian. The difference from the free-field case is that the energy spectrum is time-dependent. The integration variables of this path integral are the Grassmann functions $b, b^*, c, c^*$. They can be regarded, in a way, as the classical counterparts of the creation/annihilation operators of (anti)particles. Despite the nota-

## 21. Chiral Gauge Theories

tion, the Grassmann variables $b^*, c^*$ are not the complex conjugates of $b, c$. This is due to the asymmetry in the initial conditions: $b^*, c^*$ are fixed at $t = T$ while $b, c$ at $t = 0$. Consequently, as $b(T), c(T)$ and $b^*(0), c^*(0)$ are arbitrary, they have to be included in the path-integral measure asymmetrically:

$$d\mu(b^*, c^*; b, c) = \prod_E db_E(T) \left[ \prod_{i=1}^{(N-1)} db_E^*(t_i) db_E(t_i) \right] db_E^*(0) \times (b^* \to c^*; b \to c). \quad (1)$$

Next, I change the variables to the Grassmann fields which are the classical counterparts of the fermion field operators, $\psi^\dagger, \psi$, to get the more familiar form of the path integral. This form is the one used to derive the Feynman rules, for example. The equations connecting the two kinds of variables are just the expansion expressions for the fermion wavefunctions encountered previously.

Owing to the asymmetry of the integration measure, the associated Jacobian of this transformation is not trivial but

$$J = \det[\int d\vec{x}\, \phi_E^\dagger(\vec{x}, 0) \phi_{E'}(\vec{x}, T)] \\ \times \det[\int d\vec{x}\, \phi_{-E}^\dagger(\vec{x}, T) \phi_{-E'}(\vec{x}, 0)]; \quad E, E' > 0 \quad (2)$$

as, from the orthogonality of the basis, there is no contribution from the time slices where $b^*, c^*$ pair up with $b, c$ respectively.

Consequently, the path integral action is not just the classical action (with Grassmannian variables); the Jacobian also contributes to it. It is precisely the non-single-valuedness of the chosen basis $\{\phi_E\}$ that gives rise to the non-trivial potential-dependent Jacobian in the case of chiral gauge theories.

I now make a digression to the quantum holonomy phase of Berry [9]. For certain class of quantum mechanical hamiltonians depending on some external parameters, an eigenstate may pick up a non-dynamical phase, due to the geometry of the parameter space, after completing adiabatically a circuit in this space. The way the state traverses the circuit is governed by the Schrödinger equation. The non-trivial geometry of the parameter space is induced by the degeneracies of the hamiltonian.

The governing equation for the time development of the function $\phi_E$ can be deduced and presented in the form of a Schrödinger equation [5]. The corresponding hamiltonian is nothing more than the previous first-quantized hamiltonian shifted by the eigenvalue. Such hamiltonians admit the Berry phase [10]. With these phases, denoted by $\gamma_E$ for the energy level $E$, the correct action in the path integral representing the vacuum-to-vacuum amplitude is

$$S_{total} = S_{classical} + \sum_E sign(E)\gamma_E. \tag{3}$$

This extra contribution can be evaluated [10], upto gauge-invariant and local gauge-non-invariant terms, to be the WZ contribution cancelling the anomalous effect of the fermion part.

Without the anomaly of large gauge transformations [11], an even number of flavours would not be required for the theory with the group $SU(2)$, in particular. For small gauge transformations, the absence of anomaly preserves the first-class nature of the Gauss law constraint. That is, there is no Schwinger term in the algebra of the gauge generators as can be deduced directly from, for example, ref. [4]. An explicit construction in two dimensions [12] has also reached the same conclusion for the Gauss law constraint operators.

This gauge invariance result has also been proposed both in an *ad hoc* manner [13] and by the *formal* use of the Faddeev-Popov trick [14], the insertion of the resolution of unity into the naive path integral.

If one prefers to decompose the wavefunctions in the plane-wave basis rather than in the basis $\{\phi_E\}$ above then the corresponding Jacobian will be trivial, unlike the result in expression (2). The triviality is easily understood as the kinetic part of the first-quantised hamiltonian, the part which admits the plane-wave eigenstates, is independent of the gauge fields. Nevertheless, the gauge invariant result for the vacuum generating functional can still be obtained [15].

Although the properties of Lorentz invariance, positivity and perhaps locality (with the introduction of some new scalar field, say) are manifest, the perturbative renormalisability with the addition of the WZ term is as yet unclear. On the other hand, failure of perturbative renormalization may not be of fundamental importance if the existence of some non-trivial RG fixed point could be realised.

As there is no anomaly for chiral gauge theories, the lattice formulations of chiral fermions need some revision [16] to avoid the doubling phenomenon [17].

The method presented here could also be generalised to the second quantisation of those theories whose first-quantized hamiltonians admit Berry phase in general and to certain other kinds of anomalous field theories in particular.

## Acknowledgements

I want to thank Peter Higgs, Richard Kenway, Brian Pendleton, Diptiman Sen, Steve de Souza and David Wallace for discussions.

## References

[1] R. Jackiw, in *Anomalies, Geometry and Topology*, edited by A. White, World Scientific, Singapore, 1985; *MIT Report No.* CTP#1436, 1986, *to be published*; L. Alvarez-Gaume, *Harvard report* HUTP-85/A092, 1985.

[2] L. Faddeev, *Phys. Lett.* **145B** (1984) 81; B. Zumino, *Nucl. Phys.* **B253** (1985) 477.

[3] S.-G. Jo, *Phys. Rev.* **D35** (1987) 3179.

[4] T.D. Kieu, *Edinburgh University report* 88/441, 1988 and references therein.

[5] T.D. Kieu, *Phys. Lett.* **218B** (1989) 221.

[6] J. Wess and B. Zumino, *Phys. Lett.* **37B** (1971) 95.

[7] Y. Ohnuki and T. Kashiwa, *Prog. Theo. Phys.* **60** (1978) 548.

[8] F.A. Berezin, *The Method of Second Quantisation*, Pergamon, New York, 1966; L.D. Faddeev, in *Methods in Field Theory*, edited by R. Balian and J. Zinn-Justin, North-Holland, Amsterdam, 1975; L.D. Faddeev and A.A. Slavnov, *Gauge Fields: Introduction to Quantum Theory*, Benjamin/Cummings, Massachusetts, 1980.

[9] M.V. Berry, *Proc. Roy. Soc.* **A392** (1984) 45; B. Simon, *Phys. Rev. Lett.* **51** (1983) 2167.

[10] P. Nelson and L. Alvarez-Gaume, *Commun. Math. Phys.* **99** (1985) 103; A.J. Niemi and Semenoff, *Phys. Rev. Lett.* **55** (1985) 927.

[11] E. Witten, *Phys. Lett.* **117B** (1982) 324.

[12] G.W. Semenoff, *Phys. Rev. Lett.* **60** (1988) 680.

[13] E. d'Hoker and E. Farhi, *Nucl. Phys.* **B248** (1984) 59, 77; A.A. Andrianov and Y. Novozhilov, *Phys. Lett.* **163B** (1985) 189; L.D. Faddeev and S.L. Shatashvili, *Phys. Lett.* **167B** (1986) 225; A.J. Niemi and G.W. Semenoff, *Phys. Rev. Lett.* **56** (1986) 1019.

[14] O. Babelon, F.A. Schaposnik and C.M. Viallet, *Phys. Lett.* **177B**, (1986) 385; K. Harada and I. Tsutsui, *Phys. Lett.* **183B** (1987) 311.

[15] T.D. Kieu, *Oxford University Theoretical Physics preprint* OUTP-89-14P, 1989; K. Odaka and T. Itoh, *Lett. Math. Phys.* **15** (1988) 297.

[16] For recent works on lattice chiral gauge theories see:
K. Funakubo and T. Kashiwa, *Phys. Rev. Lett.* **60** (1988) 2113; *Phys. Rev.* **D38** (1988) 2602; S. Aoki, *Phys. Rev. Lett.* **60** (1988) 2109; *Phys. Rev.* **D38** (1988) 618.
See also T.D. Kieu, D. Sen and S.-S. Xue, *Phys. Rev. Lett.* **61** (1988) 282.

[17] For a review see, for example, T.D. Kieu, in *Quantum Field Theory as an Interdisciplinary Basis*, edited by Kunstatter et al., World Scientific, Singapore, 1988 and references therein.

DEPARTMENT OF PHYSICS, EDINBURGH EH9 3JZ AND
DEPARTMENT OF THEORETICAL PHYSICS, OXFORD OX1 3NP, U.K.

# 22

# The Correct Significance of the Binary Pulsar Observations

*D. F. Roscoe*

Virtually all current efforts directed at the unification of the gravitational force with the other forces of modern physics make the a priori assumption that the gravitational force arises exclusively from spin 2 particle exchange and, in particular, that concepts of scalar gravitation cannot be considered viable. The binary pulsar observations, Taylor et al [1], are frequently cited as conclusive evidence in support of this perspective. However, these conclusions appear to be based upon uncritical thinking; in the following, we review the analysis used, and give a statement of the correct conclusion which should be drawn from the binary pulsar experience.

Fundamentally, the binary pulsar observations consist of measurements of the arrival times of certain signals, and the best model to explain their properties appears to be provided by a binary star system having certain extreme characteristics. These arrival time measurements indicate that the system is decaying, and so one has a need to explain the mode of this decay; there are four obvious alternatives:-

- Electromagnetic radiation.
- Tidal effects.
- Mass ejection.
- Gravitational radiation.

The available information appears to discredit the viability of the first three of these mechanisms, which leaves only the fourth. Assuming that gravitational radiation is the cause of decay, then it can be modelled as a classical radiation system; that is, as *monopole effects + dipole effects + quadropole effects + ...* radiating from a spherically symmetric point source; one then asks to what extent such a model can account for the binary pulsar measurements. As Will [2] shows, the observed effects put a very low limit on the magnitude of the *dipole* component, and strongly suggest its total absence.

Any given theory of gravitation is then tested by interpreting it, where possible, as a multipole radiation model and checking its predictions vis-a-vis dipole radiation. A theory is admissible, according to the binary pulsar observations, if its dipole components are sufficiently small. The modern cliche, which states that scalar gravitation is inadmissible, arises in the following way:- there is a gravitation theory, called the Brans-Dicke theory, which consists of General Relativity + Scalar Field; if this theory is analysed as a multipole model, it is found that GR contributes exclusively to the quadropole component, whilst the Scalar Field contributes exclusively to the *dipole* component. This result has been widely mis-interpreted to mean that gravitation cannot have any scalar component; however, the correct conclusion is ... *when interpreted using a GR + Scalar Field model, the binary pulsar observations indicate that the Scalar Field component must be nearly vanishing.* That is, GR is consistent with the binary pulsar observations, but any theory consisting of GR + Scalar Field is not. This, of course, is quite *distinct* from the statement that the binary pulsar observations exclude the possibility of gravitation being a scalar effect.

To summarize, the primary particular requirement imposed by the binary pulsar on any theory is that, when interpreted as a radiation model, a multipole expansion of the model solution must contain no *dipole* component.

### References

[1] J.H. Taylor and P.M. McCulloch, Evidence for the Existence of Gravitational Radiation from Measurements of the Binary Pulsar PSR 1913 + 16, *Ann. N.Y. Acad. Sci.* **336** 442-6.

[2] C.M. Will, Theory and Experiment in Gravitational Physics, C.U.P., 1981.

DEPARTMENT OF APPLIED MATHEMATICS, UNIVERSITY OF SHEFFIELD,
SHEFFIELD, ENGLAND.

# 23

## Approaches to Scale Invariance in Two Dimensional Statistical Mechanics

*P.P. Martin*

The field theory limit of critical $q = 4\cos^2(\pi/r)$ $(r \in Z)$ state Potts and related models is a Conformal Field Theory (CFT) with central charge $c = 1 - 6/r(r-1)$. This may be shown by considering charged Coulomb gas models which are known to be universal with the Potts models. The arguments, however, are tortuous [1]. We find vestigial properties in the Potts models associated with the CFT limit directly by means of block spin transformations. These are motivated by cabling transformations of braids, which arise in the transfer matrix formulation of such models.

We will first discuss the notion of a transfer matrix. Consider a set of curves on a closed surface (other than $S_2$) constructed as follows. Chop the surface into 3-punctured spheres ('pants') and empty discs. For each pants draw 2 generating elements of the fundamental group. Draw as many non-touching circles as you like around each puncture. These must not touch the generators. Draw as many non-touching lines as you like (almost) between each pair of punctures. You must choose the total number of lines at a puncture to match its partner in the original surface. Sewing up the original surface by identifying lines across boundaries allows some ambiguity (see later). All crossings have coordination number 4 and all faces are quadrilateral, except for 2 hexagons per pants (there is a dual construction giving all quadrilateral faces). Arcs of curves between crossings will be called edges.

Now consider spins $s_x$, each taking values from some set $V$, on the crossings (labelled $x$), and interactions coupling between spins on the edges (labelled $i$). There exist models with various permutations of these roles for the simplices. The principles are the same, our choice is arbitrary. Draw a closed curve or curves [c] on the surface passing through crossings but not edges. Consider the spins inside and on [c] (you may have to decide which submanifold is inside). Call these interior and exterior spins respectively. The partition vector $Z[c]$ is a vector in the space of possible configurations of the exterior spins (we will also call this space [c] in general). The elements give the partition functions for the appropriately bounded interior system, ie. the Boltzmann weights for the interior interactions summed over the interior spin configurations. The partition vector for a pants with punctures designated $[c:l_1, l_2, l_3]$ may be written

$$Z[c:l_1,l_2,l_3] = \sum_{i_1 i_2 i_3} (P_{(\underline{a})})_{i_1 i_2 i_3} (T_{(a_1)}{}^{b_1})_{i_1 l_1} (T_{(a_2)}{}^{b_2})_{i_2 l_2} (T_{(a_3)}{}^{b_3})_{i_3 l_3}, \quad (1)$$

where $(P_{(\underline{a})})_{i_1 i_2 i_3}$ is the partition vector for a pants with only one circle per puncture; $i_1, i_2, i_3$ give the exterior configuration for each puncture; $\underline{a} = (a_1, a_2, a_3)$ is the number of lines per puncture and $b_1, b_2, b_3$ the remaining circles per puncture. The object $T_{(a)}$ is the a-site cylindrical layer transfer matrix on $\otimes^a V$. Punctures may be sewn together by matrix multiplication $..T_{(a)}(G_{(a)})^{2p} T_{(a)}..$ ($p = 0, 1, .., a-1$) where the matrix $G^{2p}$ determines the identification of lines at the boundary.

Consider the transfer matrix for adding a single interaction $i$ to the partition vector $Z[c]$. That is to say, for including an interaction between $s_x$ and $s_{x+1}$ adjacent on [c]; or between $s_x$ and $s_{x'}$ internalising $x$ and thus modifying [c] to [c']. This transfer matrix may be written in the form:

$$t_i(y) = a(y)1 + b(y)g_i + c(y)h_i + .., \quad (2)$$

where $a(y), b(y), c(y),..$ are scalar functions of the coupling strengths (generically $y$). If a spin is internalised the matrices $1, g, h,..$ map $V_x \to V_{x'}$ ; if not they are diagonal on $V_x \otimes V_{x+1}$ . In any case they span some $End(V)$ in general. In most cases the symmetries of the system simplify this. The number of summands in (2) equals the number of distinct possibilities for the interaction energy. Thus, for example, the Potts model has

$$t_i(y) = a(y)1 + b(y)g_i. \quad (3)$$

In the bulk of the system we will label edges on circles with even numbers and those on lines with odd. Any $t_i$ that does not change the boundary [c] only changes the coupling , so $\exists\ f(y)$ such that $t_i^2(y) = t_i(f(y))$ and, using (3)

$$g_i^2 + \beta g_i + \alpha 1 = 0, \quad (4)$$

where $\alpha$ and $\beta$ are constants (for the $4\cos^2(\pi/r)$ state model with $Q = -\exp(\pi i/r)$ we have $\alpha = -Q^2$, $\beta = -1 + Q^2$). The Potts models are part of a large class which satisfy the star-triangle relations in some one-dimensional submanifold of coupling parameter space, ie. for which $\exists\ y''(y,y')$ such that

$$t_i(y)t_{i+1}(y')t_i(y'') = t_{i+1}(y'')t_i(y')t_{i+1}(y), \tag{5}$$

where edge $i+1$ has a site in common with edge $i$. The Statistical Mechanical significance of this is that it means that transfer matrices for modifying an extended arc of [c] commute (up to boundary terms), even with different values of the coupling parameter. This can be a powerful computational aid [2]. It is usually possible to choose $a(y), b(y)$ such that

$$g_i g_{i+1} g_i = g_{i+1} g_i g_{i+1}, \tag{6}$$

so each such model gives representations of A-type Hecke algebras [3]. For example, on the spins round a circle $\otimes^a V$ we can build $\{g_i; i = 1,..,2a-1\}$ corresponding to $a$ edges on lines and $a-1$ edges round the circle ($g_{2a}$ is not independent). Up to degeneracy the algebra then determines the correlation functions of the model. The puncture repair kit $G$ is essentially

$$G_{(a)} = \prod_{m=1}^{2a-1} g_m, \tag{7}$$

since $G_{(a)} g_i (G_{(a)})^{-1} = g_{i+1}$, and

$$H_{(a)} = \prod_{l=1}^{2a-1} \prod_{m=1}^{2a-l} g_m \tag{8}$$

gives the reflection $H_{(a)} g_i (H_{(a)})^{-1} = g_{2a-i}$. Note that $H_{(a)}^2 = G_{(a)}^{2a}$ is central.

These algebras have a known structure for generic $q$, which breaks down for each $q = 4\cos^2(\pi/r)$ to leave an algebra with a finite number of inequivalent irreducible unitarisable representations. Similarities with the Virasoro algebra can be explored directly at this level, or by examining the response of lattice models to conformal transformations in some approximation. In what follows we look at the response to scale changes.

Consider replacing each curve on the surface by $n$ close but not touching curves. The number of sites is increased by a factor of $n^2$. We want to think of this as giving a more detailed description of the same physical system. That is, the physical edge length is reduced. We expect to have to retune the coupling parameters (and turn on some new ones) to keep this description pertinent.

In the simplest scenario for block spin renormalisation we invert this process in order to compute the retuning. We have a partition function written in terms of the Hamiltonian for some set of fields (spins resolved at some scale $\mu$) with various coupling parameters $\{y\}_\mu$, and the same partition function written in terms of a cruder set of fields resolved at $\nu\mu$ ($\nu > 1$) with couplings $\{y(\{y\}_\mu)\}_{\nu\mu}$. In either process the fixed points of the implied coupling transformations are associated with scale invariant critical points. Because we have not turned on enough transfer matrix technology to allow arbitrary interactions on the lattice we will remain in the edge coupling subspace. We will sit on a critical point and allow the spins (through (2) or (3)) rather than the couplings to be transformed. Any fixed point will be identified as a fixed point of (4) and (6).

The critical points in the edge coupling subspace are known. They include the point $y'' = y' = y$, $a(y) = 0$, $b(y) = 1$ at which all our transfer matrices become braids (with quotient relations (4)). Note, from (5), that apart from the parameters $y$ the $\{t_i\}$ themselves behave like braids. The obvious blocking transformation, then, just corresponds to a cabling of braids, so (6) is automatically preserved under the change

$$t_i^{(n)} = (\prod_{k=1}^{n} \prod_{j=1-k}^{k-1}{}' t_{ni+j})(\prod_{k=n-1}^{1} \prod_{j=1-k}^{k-1}{}' t_{ni+j}), \qquad (9)$$

(where $\prod_j'$ is incremented in steps of 2) at the given critical point. In (9) $n$ strings crossing $n$ strings is read as one crossing, or $n^2$ interactions are replaced by one.

It is easy to prove that (5) as well as (6) is preserved by the cabling (9), so if we have a solution of (5) all the block spin descendants also give solutions [4]!

Without some (approximate) decimation procedure it is not generally possible to preserve the form of (3) under (9). Even if we work at the critical point the descendant relation is not consistent with (4). This is just as well, since most of these models do not have scale invariant limits! The exceptions are the cases $q = 4\cos^2(\pi/r)$, for which the appropriate quadratic relation can appear as a factor in a higher order quotient relation for the descendant operators. In other words we are generating larger algebras, but in these special cases their relations are consistent with the existence of quotients isomorphic to the original algebras with the number of generators rescaled (and thus with the same spectrum as the original models on a cruder lattice). We have to check that these quotients are represented in the matrices we build from (9). A toy example is the Burau representation [4] for $nN$ generators $R_{B_{nN}}(g_i(Q))$

$$(R_{B_{nN}}(g_i(Q)))_{jk} = \delta_{j,k}(1 - \delta_{i,k}(1 - Q^2)) + Q\delta_{j,i}(\delta_{k+1,i} + \delta_{k-1,i}), \qquad (10)$$

for which we find the relevant decomposition

$$R_{B_{nN}}(g_i^{(n)}(Q)) = R_{B_N}(g_i(Q^n)) \oplus ... \qquad (11)$$

In other words this reducible representation of some larger algebra contains an irreducible component which gives the Burau representation of the original algebra again provided $q = 4\cos^2(2\pi k/(n-1))$ ($k \in Z$).

Another interesting example is the case $q = 4$, for which the quadratic relation is manifestly preserved under (9). In general, at least for simple topologies, the identification of irreducible representations with particular parts of the spectrum (the free energy, long distance correlations etc.) can be made in a scale independent way. It remains to check that these are preserved. Here $\{t_i^{(n)}\}$ generates a subgroup $S_N$ in the symmetric group $S_{nN}$. A conjugate realisation of this subgroup with an easier restriction procedure is indicated by $S_{nN} \supset \otimes^n S_N \supset diagonal\ S_N$ [6]. We find that the representation associated with the free energy, for instance, is generally preserved. In the large lattice limit this corresponds to scale invariance. Note that any transformation corresponding to a conjugate realisation of the subgroup has a similar invariance.

To summarise: we find that the response of critical $q$-state Potts models to $n$-fold cabling like block spin transformations is to alter the model in a way which depends on $q$. However the models with $q = 4\cos^2(\pi/r)$ ($r \in Z$) may retain an invariant sector, the extent of which depends on $r$ and $n$.

### References

[1] see, for example, B. Nienhuis in *Phase Transitions and Critical Phenomena*, eds. C. Domb and J.L. Lebowitz, vol. 11; and J. Cardy, ibid.

[2] R.J. Baxter, *Exactly Solved Models in Statistical Mechanics*, Academic Press, New York, 1982.

[3] P.N. Hoefsmit, *Representation Theory of Hecke Algebras*, University of British Columbia thesis, 1974.

[4] see also E. Date, M. Jimbo, A. Kuniba, T. Miwa and M. Okado, *Advanced Studies in Pure Math.* **16** (1988) 17; *Nucl. Phys.* **B290[FS20]** (1987) 231.

[5] J.S. Birman, *Ann Math Studies* **82** (1974) 1.

[6] G. Robinson, *Representation Theory of the Symmetric Group*, University Press, Edinburgh, 1961.

DEPARTMENT OF MATHEMATICS, BIRMINGHAM UNIVERSITY, BIRMINGHAM, ENGLAND.

# 24

## String Amplitudes and Twistor Diagrams: an Analogy

*A. P. Hodges*

A proposal has been made [1,2,3] for a conformal field theory in four dimensions (CFT4), in which the *twistor* representation of quantum fields plays an essential role. According to this proposal, a scattering amplitude could arise in the first instance as associated with a specific complex manifold $X$ (bearing a specified relationship to flat twistor space); to obtain a physically meaningful amplitude a summation of these amplitudes would be performed over (a class of) such manifolds. It was noted that physically meaningful amplitudes (for theories of interacting massless fields in Minkowski space-time) have already appeared in twistor theory. They have arisen in the formalism of *twistor diagrams* [4,5,6] — compact contour integrals in products of twistor spaces, with a rough analogy to Feynman diagrams, but with certain features similar to the dual diagrams of bosonic string theory. It was therefore suggested that there might exist a direct connection between the CFT4 picture and the twistor diagram formalism. This speculation is here strengthened by noting an analogy between standard two-dimensional string theory and a specific twistor-diagrammatic calculation, suggesting that the integration of twistor diagrams could be interpreted as a summation over complex manifolds.

We consider the derivation of the Veneziano amplitude for four open strings of spin 0. According to the standard theory [7, page 49], the amplitude associated with one string is determined by mapping that string conformally onto a disc with four boundary points removed, or equivalently, the upper-half-plane with four real points removed. We shall use the latter formulation. The Veneziano amplitude then results from summing over the amplitudes associated with all such punctured half-planes.

Such half-planes can be labelled by four real points $x_1, x_2, x_3, x_4$. But two such half-planes are conformally equivalent if these parameters have the same cross-ratio. Thus to count each manifold just once, the summation should run only over values of the cross-ratio. This can be achieved by fixing $x_1, x_3, x_4$ say as $0, 1, \infty$ (a 'gauge') and then summing over $x_2$. More symmetrically we may formally sum over all four parameters and then divide by the infinite volume of the $SL(2, R)$ 'gauge group'. The formalism proposed here has the symmetry of the latter approach, but avoids infinities by replacing the original non-compact integral by the compact integration of a projective form in $(CP^1)^4$. This requires a number of steps: (1) re-interpreting the original real integral as the integration over a real path of a complex form (2) using projective spinors (appropriate in any case because this puts $\infty$ on an equal footing with other points) (3) using a Pochhammer contour to replace the integration into branch points by compact contour integration round the branch points (4) restoring the symmetry by writing this as an integral in $(CP^1)^4$.

The integral for the amplitude corresponding to the cyclic ordering (1234) (using the 'gauge' choice given above) is

$$\int_0^1 dx_2 \mid x_2 \mid^{k_1.k_2} \mid 1 - x_2 \mid^{k_2.k_3}.$$

This becomes (on following these four steps) the spinor integral

$$(2\pi i)^{-2} \oint Dz_1 \wedge Dz_2 \wedge Dz_3 \wedge Dz_4 \ [(z_1.z_3)(z_2.z_4)]^{-k_1.k_3}$$

$$\times \frac{[(z_1.z_2)(z_3.z_4)]^{-k_1.k_2}}{2i \sin \pi(k_1.k_2)} \frac{[(z_1.z_4)(z_2.z_3)]^{-k_1.k_4}}{2i \sin \pi(k_1.k_4)}$$

which we write diagrammatically as

The 'tachyonic' condition $m^2 = -k_1.k_2 - k_1.k_3 - k_1.k_4 = -2$ is equivalent to this projective integral being well-defined.

This integral may be considered as composed of singular 'propagator' factors of form

$$\frac{[(z_1.z_2)(z_3.z_4)]^{-k_1.k_2}}{2i \sin \pi(k_i.k_2)},$$

and numerator factors (actually *periods* of these singular factors) of form

$$[(z_1.z_3)(z_2.z_4)]^{-k_1.k_3}.$$

## 24. String Amplitudes and Twistor Diagrams

For general values of momenta this distinction is artificial, but when the exponents are integers (the case of interest when studying the twistor analogue) the 'propagators' become logarithmic and the 'numerators' non-singular.

If the external states have $SU(n)$ indices ('quark-antiquark charge' in the original bosonic string theory) then a further coefficient must be specified, namely $\operatorname{tr}(\Lambda_1\Lambda_2\Lambda_3\Lambda_4)$, where $\Lambda_i$ is the $SU(n)$ matrix on the $i$th string. The complete amplitude is then given by the sum:

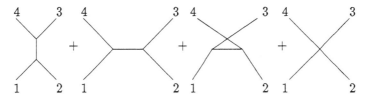

$$\operatorname{tr}(\Lambda_1\Lambda_3\Lambda_2\Lambda_4)\left|\begin{array}{c}4\text{---}3\\ \times\\ 1\text{---}2\end{array}\right| + \operatorname{tr}(\Lambda_1\Lambda_2\Lambda_4\Lambda_3)\left|\begin{array}{c}4\text{---}3\\ \times\\ 1\text{---}2\end{array}\right|$$

$$+ \operatorname{tr}(\Lambda_1\Lambda_2\Lambda_3\Lambda_4)\left|\begin{array}{c}4\text{---}3\\ \times\\ 1\text{---}2\end{array}\right| \tag{1}$$

We now turn attention to field theory in Minkowski space, in fact to pure $SU(2)$ gauge field scattering. The reason for this choice of process is that it turns out to be in a certain sense the simplest to describe as a twistor integral. We shall observe a remarkable parallel to this string-theoretic formula which emerges from this twistorial re-description. To do this we compute this amplitude by standard Feynman rules: this amounts to summing

where the $i$th state is defined by potential $\Phi_i^a$ and $SU(2)$ matrix $\Lambda_i$. It turns out that the scattering demonstrates 'helicity conservation', with left and right helicity parts interacting independently. Hence we lose nothing by taking $\Phi_1^a, \Phi_2^a$ to be self-dual, i.e. with 2-spinor representations $\Phi_1^{AA'}$, $\Phi_1^{AA'}$ satisfying

$$\nabla_{BA'}\Phi_1^{AA'} = 0, \ \nabla_{BA'}\Phi_2^{AA'} = 0,$$

and the other fields likewise to be anti-self-dual: i.e.

$$\nabla_{B'A}\Phi_3^{AA'} = 0, \ \nabla_{B'A}\Phi_4^{AA'} = 0.$$

Then the fields are given in 2-spinor form by $\phi_1^{A'B'}(x), \phi_2^{A'B'}(x), \phi_3^{AB}(x)$, $\phi_4^{AB}(x)$, with Fourier transforms $\tilde{\phi}_1^{A'B'}(k_1)$ etc. The general case, in which the interacting fields are not eigenstates of helicity, can be recovered by linearity. The individual Feynman diagrams are gauge-dependent but their sum (by a straightforward but non-trivial computation) may be expressed in the manifestly gauge-invariant form:

$$\operatorname{tr}(\Lambda_1\Lambda_3\Lambda_2\Lambda_4) \int d^4k_1 \ldots d^4k_4 \, \frac{\tilde{\phi}_1^{A'B'}(k_1)\tilde{\phi}_3^{AB}(k_3)\tilde{\phi}_{2A'B'}(k_2)\tilde{\phi}_{4AB}(k_4)}{(k_1+k_3)^2 \, (k_1+k_4)^2}$$

$$+ \operatorname{tr}(\Lambda_1\Lambda_3\Lambda_4\Lambda_2) \int d^4k_1 \ldots d^4k_4 \, \frac{\tilde{\phi}_1^{A'B'}(k_1)\tilde{\phi}_2^{AB}(k_2)\tilde{\phi}_{3A'B'}(k_3)\tilde{\phi}_{4AB}(k_4)}{(k_1+k_2)^2 \, (k_1+k_3)^2}$$

$$+ \operatorname{tr}(\Lambda_1\Lambda_2\Lambda_3\Lambda_4) \int d^4k_1 \ldots d^4k_4 \, \frac{\tilde{\phi}_1^{A'B'}(k_1)\tilde{\phi}_2^{AB}(k_2)\tilde{\phi}_{3A'B'}(k_3)\tilde{\phi}_{4AB}(k_4)}{(k_1+k_2)^2 \, (k_1+k_4)^2}$$

Using standard translation techniques (see for instance [8]), each of these terms may be translated into a twistor diagram, yielding the sum:

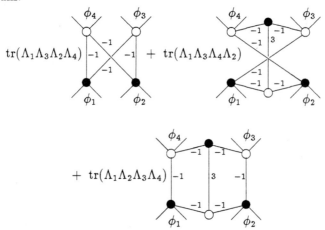

The definition of these diagrams is specified elsewhere [4,5,6] but a brief description can be given. The notation is like Feynman diagram notation, in that all the vertices represent variables to be integrated out. But these variables are twistors or dual twistors (corresponding to black or white vertices respectively) and the integration is compact contour integration. The external fields appear in the standard twistor representation, i.e. as first cohomology group elements. The 'propagator' lines connecting the vertices are simple singular factors. The whole structure is manifestly finite and is also manifestly conformally invariant. However, the aspect of the integration that concerns us here is the analogy with the sum of spinor integrals (1).

The essential point is that in each diagram the form to be integrated is simply the product of the external fields and a natural volume form on the total space. The contour over which it is to be integrated is dictated by the presence of lines labelled $(-1)$. These lines define logarithmic factors in the integral, and the contour can be regarded as a higher-dimensional Pochhammer contour which winds round the branch curves they define. Note that the corresponding 'numerators' are just unity in this case, and also that these logarithmic factors connect the external states in the same cyclic order as in the trace yielding the $SU(2)$ coefficient. Thus the sum of the twistor diagrams is of just the same form as the sum (1) of spinor integrals. We know that these spinor integrals can be derived as integrals over a parameter space of amplitudes arising from a conformal field theory. This suggests that it may be possible to derive the analogous twistor integral from a CFT4 principle, instead of producing it by translation from field theory.

The pure $SU(2)$ gauge field scattering integral is special because only for it does the twistor diagram representation reduce to the integration of a volume form, making the analogy with string theory particularly close. To describe the interaction of fields with spin other than 1, one must integrate certain very simple rational functions rather than pure volume. However, there seems no reason why a generalisation encompassing this feature could not follow from a CFT4 principle.

### References

[1] A. P. Hodges, R. Penrose and M. A. Singer, *Phys. Letters* **B216** (1989) 48–52.

[2] R. Penrose, in *this volume*.

[3] M. A. Singer, in *this volume*.

[4] A. P. Hodges, *Physica* **114A** (1982) 157–175.

[5] A. P. Hodges, *Proc. R. Soc. Lond.* **A 397** (1985) 341–374.

[6] A. P. Hodges, *Proc. R. Soc. Lond.* **A 397** (1985) 375–396.

[7] M. B. Green, J. H. Schwarz and E. Witten, *Superstring theory*, Cambridge University Press, Cambridge, 1987.

[8] A. P. Hodges, *Proc. R. Soc. Lond.* **A 386** (1983) 185-210.

MATHEMATICAL INSTITUTE, 24-29 ST. GILES', OXFORD OX1 3LB, ENGLAND.

# 25

## Lie Cochains on an Algebra

*Jacek Brodzki*

The aim of this note is to describe a relation between Hochschild cochains and Lie cochains that can be defined on an algebra $A$. This paper serves as a starting point for further investigations of geometric interpretation of cyclic cohomology.

We first recall the basic properties of the bar construction of an algebra presented in [2] and [3]. Let $A$ be an algebra with unit. Let us consider the tensor coalgebra $B(A) = T(A[1])$ with $A$ located in degree one with the coproduct given by

$$\Delta(a_1, \ldots, a_p) = \sum_{i=0}^{p} (a_1, \ldots, a_i) \otimes (a_{i+1}, \ldots, a_p) \qquad (1)$$

for $(a_1, \ldots, a_p) \in A^{\otimes p}$. There is a counit $\eta$, i.e. a map $\eta : B(A) \to k$, which is the projection onto $A^{\otimes 0} = k$. The coalgebra $B(A)$ can be made into a DG coalgebra by introducing a degree $-1$ differential which on $B(A)_p = A^{\otimes p}$ is given by the formula

$$b'(a_1, \ldots, a_p) = \sum_{i=1}^{p-1} (-1)^{i-1}(a_1, \ldots, a_i a_{i+1}, \ldots, a_p). \qquad (2)$$

The coalgebra $B(A)$ equipped with the differential $b'$ is called the bar construction of the algebra $A$. If $V$ is a vector space then the space of cochains $\mathrm{Hom}(B(A), V)$ can be given a differential $\delta$ of degree 1. For $f \in \mathrm{Hom}(B(A)_p, V)$ we define $\delta f = (-1)^{p+1} f b'$. If $R$ is an algebra then $\mathrm{Hom}(B(A), R)$ has the structure of a differential graded algebra. If $f$ and $g$ are $p$ and $q$ cochains respectively, then their product is given by

$$(fg)(a_1, \ldots, a_{p+1}) = (-1)^{pq} f(a_1, \ldots, a_p) \cdot g(a_{p+1}, \ldots, a_{p+q}). \qquad (3)$$

A degree $-1$ map $\rho : B(A) \to R$ is called a twisting cochain if it satisfies $\delta \rho + \rho^2 = 0$. An example of a twisting cochain is the canonical map $\theta : B(A) \to A$. Let now $M$ be an $A$-bimodule. The space $\mathrm{Hom}(B(A), M)$ of cochains with values in $M$ is a DG-bimodule over $\mathrm{Hom}(B(A), A)$ so one can define a degree one map

$$(\delta + \mathrm{ad}\,\theta)f = \delta f + \theta f - (-1)^{|f|} f \theta. \qquad (4)$$

One checks that the twisting cochain property of $\theta$ insures that this map is a differential, i.e. $(\delta + \mathrm{ad}\theta)^2 = 0$. The elements of the space $\mathrm{Hom}(B(A), M)$ with the twisted differential $\delta + \mathrm{ad}\theta$ are Hochschild cochains with values in a bimodule $M$.

The algebra $A$ can be given a Lie algebra structure by defining a bracket $[a, b] = ab - ba$; $a, b \in A$. This Lie algebra will be denoted $\mathrm{Lie}(A)$. There are two graded vector spaces associated with $\mathrm{Lie}(A)$. One is the space of Lie chains $C_*(\mathrm{Lie}(A))$, which is the same as the exterior algebra of $\mathrm{Lie}(A)$, the other is the space $C^*(\mathrm{Lie}(A), M) = \mathrm{Hom}(C_*(\mathrm{Lie}(A)), M)$ of Lie cochains with values in $M$ which is regarded here as a $\mathrm{Lie}(A)$-module with the adjoint action of $\mathrm{Lie}(A)$. One can define a differential $d$ on $C^*$ which for a Lie $p$-cochain $g$ is given by the formula

$$dg(a_1 \wedge .. \wedge a_{p+1}) =$$
$$(-1)^{p+1}\Big\{\sum_i (-1)^i [a_i, g(a_1 \wedge \ldots \wedge \hat{a}_i \wedge \ldots \wedge a_{p+1})] \qquad (5)$$
$$+ \sum_{i<j} (-1)^{i+j+1} g([a_i, a_j] \wedge a_1 \wedge .. \wedge \hat{a}_i \wedge .. \wedge \hat{a}_j \wedge .. \wedge a_{p+1})\Big\}.$$

This formula differs by a sign $(-1)^{p+1}$ from a classical one. The sign appears here to make a construction compatible with the sign rule introduced in the definition of the differential $\delta$.

We now want to define a map $\mathrm{Hom}(B(A), M) \to C^*(\mathrm{Lie}(A), M)$. Given a Hochschild $p$-cochain $f$ one defines a corresponding Lie cochain $\hat{f}$ by

$$\hat{f}(a_1 \wedge .. \wedge a_p) = \mathcal{A}f(a_1, .., a_p) = \sum_{\sigma \in \Sigma_p} \mathrm{sgn}(\sigma) f(a_{\sigma 1}, .., a_{\sigma p}). \qquad (6)$$

where $\mathcal{A}$ denotes the alternation operation. In this way the graded vector space $C^*(\mathrm{Lie}(A), M)$ is a quotient of $\mathrm{Hom}(B(A), M)$.

**Lemma.** *The twisted differential $\delta + \mathrm{ad}\theta$ defined in $\mathrm{Hom}(B, M)$ induces via the map $\mathcal{A}$ the differential $d$ on the space $C^*(\mathrm{Lie}(A), R)$ of Lie cochains.*

Proof. One calculates

$$\mathcal{A}fb'(a_1, \ldots, a_{p+1})$$
$$= \sum_{\sigma \in \Sigma_{p+1}} \sum_{i=1}^{p} (-1)^{i-1} f(a_{\sigma 1}, \ldots, a_{\sigma i} a_{\sigma(i+1)}, \ldots, a_{\sigma(p+1)})$$
$$= \frac{p(p+1)}{2} \sum_{\substack{\sigma \in \Sigma_{p+1} \\ \sigma 1 < \sigma 2}} f([a_{\sigma 1}, a_{\sigma 2}], \ldots, a_{\sigma(p+1)})$$

## 25. Lie Cochains on an Algebra

$$= \sum_{i<j}(-1)^{i+j+1}\hat{f}([a_i,a_j]\wedge\ldots\wedge\hat{a}_i\wedge\ldots\wedge\hat{a}_j\wedge\ldots\wedge a_{p+1})$$

and analogously

$$\mathcal{A}\operatorname{ad}\theta f(a_1,\ldots,a_{p+1})$$
$$=(-1)^p\sum(-1)^{i-1}[a_i,\hat{f}(a_1\wedge\ldots\wedge\hat{a}_i\wedge\ldots\wedge a_{p+1})]$$

The above formulae combine to give the differential $d$ introduced in (5).

Let g be the algebra of $n\times n$ scalar matrices. Then the algebra $M_n(A)$ of matrices with entries in $A$ is isomorphic to $A\otimes\mathbf{g}$. As in the case of the algebra $A$ one forms the space $\operatorname{Hom}(B(A\otimes\mathbf{g}),M\otimes\mathbf{g})$ of cochains with values in $M\otimes\mathbf{g}$. We shall give a more detailed description of the space of matrix valued cochains. First we note that the $\operatorname{Hom}(B(A),A)$-module structure can be extended to matrix cochains. This is clear if we write

$$\operatorname{Hom}((A\otimes\mathbf{g})^{\otimes p},M\otimes\mathbf{g})=\operatorname{Hom}(A^{\otimes p},M)\otimes\operatorname{Hom}(\mathbf{g}^{\otimes p},\mathbf{g}). \quad (7)$$

and require that a cochain $\theta$ acts as $\tilde{\theta}=\theta\otimes 1$. This enables one to equip the space of matrix cochains with a twisted differential $\delta+\operatorname{ad}\tilde{\theta}$.

The space of Hochschild cochains sits inside the space of matrix cochains. To see this we define a map

$$\operatorname{Hom}(B(A),M)\longrightarrow\operatorname{Hom}(B(A\otimes\mathbf{g}),M\otimes\mathbf{g}) \quad (8)$$

which for $f\in\operatorname{Hom}(B(A)_p,M)$ and $X^i\in\mathbf{g}$ is given by

$$\tilde{f}(a_1\otimes X^1,\ldots,a_p\otimes X^p)=f(a_1,\ldots,a_p)\otimes X^1\cdots X^p \quad (9)$$

We denote the Lie algebra $\mathbf{gl}_n(A)=\operatorname{Lie}(A\otimes\mathbf{g})$ associated to the algebra $A\otimes\mathbf{g}$ by the same symbol. It is clear from (7) that the space of matrix cochains has a structure of a g-module, where g stands for the Lie algebra of $n\times n$ scalar matrices. Our next goal is to identify the subspace invariant under this action. To this end we write

$$\operatorname{Hom}((A\otimes\mathbf{g})^{\otimes p},M\otimes\mathbf{g})^{\mathbf{g}}=\operatorname{Hom}(A^{\otimes p},M)\otimes(\operatorname{Hom}(\mathbf{g}^{\otimes p},\mathbf{g}))^{\mathbf{g}}. \quad (10)$$

From the invariant theory for $\mathbf{gl}_n$ follows that if $n>p$ the space $\operatorname{Hom}(\mathbf{g}^{\otimes p},\mathbf{g})^{\mathbf{g}}$ may be given a basis $\{\Phi_\pi\}$ of functionals, indexed by elements of the symmetric group $\Sigma_{p+1}$, which we describe as follows. Let $\pi$ be a permutation of $\{0,1,\ldots,p\}$ which decomposes into a product of disjoint cycles as $\pi=\sigma_0\sigma_1\cdots\sigma_n$, where $\sigma_0$ is the cycle containing zero. If $\sigma$ is the cycle $(i_1,\ldots,i_l)$ put $X^{(\sigma)}=X^{i_1}\cdots X^{i_l}$. Given such a permutation $\pi$ we define a functional $\Phi_\pi$ by ($X^0=1$)

$$\Phi_\pi(X^1,\ldots,X^p)=X^{(\sigma_0)}\prod_{i\geq 1}\operatorname{tr}X^{(\sigma_i)} \quad (11)$$

By invariant theory the basis $\{\Phi_\pi\}$ gives the following identification

$$\mathrm{Hom}(\mathrm{g}^{\otimes p}, \mathrm{g})^{\mathrm{g}} = k[\Sigma_{p+1}] \tag{12}$$

In the special case where $\pi$ is the cycle $(0, 1, \ldots, p)$ we see that the space of invariant matrix cochains includes all cochains of the form $f(a_1, \ldots, a_p) \otimes X^1 \cdots X^p$ for some Hochschild $p$-cochain $f$. But on the other hand cochains of that form constitute the image of the map $f \mapsto \tilde{f}$ defined in (9). Hence the space of Hochschild cochains $\mathrm{Hom}(B(A), M)$ is included in the space of g-invariant matrix valued cochains $\mathrm{Hom}(B(A \otimes \mathrm{g}), M \otimes \mathrm{g})^{\mathrm{g}}$.

We next describe the g-invariant Lie cochains. First we note that

$$\begin{aligned} C^p(A \otimes \mathrm{g}, M \otimes \mathrm{g})^{\mathrm{g}} &= \Big(\mathrm{Hom}((sgn) \otimes (A \otimes \mathrm{g})^{\otimes p}, M \otimes \mathrm{g})\Big)^{\mathrm{g} \times \Sigma_p} \\ &= \Big(\mathrm{Hom}((sgn) \otimes A^{\otimes p}, M) \otimes k[\Sigma_{p+1}]\Big)^{\Sigma_p} \end{aligned}$$

where $(sgn)$ denotes the one dimensional sign representation of $\Sigma_p$. The group $\Sigma_p$ acts on the functionals $\Phi_\pi$ by $\sigma \cdot \Phi_\pi = \Phi_{\sigma\pi\sigma^{-1}}$ for $\sigma \in \Sigma_p$ and $\pi \in \Sigma_{p+1}$, where $\Sigma_p$ is embedded in $\Sigma_{p+1}$ as the permutations that leave 0 fixed. The orbits of this action are distinguished by $i$ ($0 \leq i \leq p$), the length of the cycle containing 0. In each orbit we choose a representative for which the cycle containing 0 is $(0, 1, \ldots, i)$. The group $\Sigma_{p-i}$ acts then on the remaining cycles. This gives the following isomorphisms

$$\begin{aligned} \Big(\mathrm{Hom}((sgn) \otimes A^{\otimes p}, M) \otimes k[\Sigma_{p+1}]\Big)^{\Sigma_p} & \\ = \bigoplus_{i=0}^{p} \Big(\mathrm{Hom}(A^{\otimes i} \otimes (sgn) \otimes A^{\otimes p-i}, M) \otimes k[\Sigma_{p-i}]\Big)^{\Sigma_{p-i}} & \\ = \bigoplus_{i=0}^{p} \mathrm{Hom}\Big(A^{\otimes i} \otimes \big\{\big((sgn) \otimes A^{\otimes p-i}\big) \otimes_{\Sigma_{p-i}} k[\Sigma_{p-i}]^*\big\}, M\Big) & \\ = \bigoplus_{i=0}^{p} \mathrm{Hom}(C_{p-i}(A \otimes \mathrm{g})_{\mathrm{g}}, \mathrm{Hom}(A^{\otimes i}, M)) & \tag{13} \end{aligned}$$

where $C_{p-i}(A \otimes \mathrm{g})_{\mathrm{g}}$ is the subspace of coinvariant Lie chains [1].

Let $i = p$. In this case the invariant Lie cochains are produced from Hochschild cochains by means of the composite

$$\mathrm{Hom}(B(A), M) \longrightarrow \mathrm{Hom}(B(A \otimes \mathrm{g}), M \otimes \mathrm{g}) \xrightarrow{\mathcal{A}} C^*(A \otimes \mathrm{g}, M \otimes \mathrm{g}) \tag{14}$$

where, as in the case of the algebra $A$, $\mathcal{A}$ denotes the alternation operation. Explicitly, if $f$ is a Hochschild $p$-cochain then the corresponding invariant Lie cochain is

$$\hat{f}(a_1 \otimes X^1 \wedge \ldots \wedge a_p \otimes X^p) = \sum_{\sigma \in \Sigma_p} sgn(\sigma) f(a_{\sigma 1}, \ldots, a_{\sigma p}) \otimes X^{\sigma 1} \cdots X^{\sigma p}. \tag{15}$$

This means that the space $\mathrm{Hom}(B(A), M)$ is included in the space of invariant Lie cochains as the $i = p$ subspace in the decomposition (13).

We can summarize our considerations with the following

**Theorem.** *In degrees* $p < n = \dim \mathfrak{g}$ *one has*

$$C^*(A \otimes \mathfrak{g}, M \otimes \mathfrak{g})^{\mathfrak{g}} = \mathrm{Hom}(C_*(A \otimes \mathfrak{g})_{\mathfrak{g}}, \mathrm{Hom}(B(A), M))$$

Proof. The proof of the theorem will be complete when we check that the isomorphism respects the differentials. Let us fix $i, j$, $i + j = p$ and choose a cochain $f \in \mathrm{Hom}(C_i(A \otimes \mathfrak{g})_{\mathfrak{g}}, \mathrm{Hom}(A^{\otimes j}, M))$. If we refer again to the given above identification of the coinvariant Lie chains we can write $f(a_1, \ldots, a_i)(a_{i+1}, \ldots, a_p) \in M$ for $a_k \in A$. This expression is skew-symmetric in the first $i$ variables. If we skew-symmetrize it in the remaining $j$ variables and then perform shuffle alternation mixing the two sets of $a$'s we obtain a skew-symmetric functional $\hat{f} \in \mathrm{Hom}(C_p(A), M)$. We then define the corresponding Lie cochain $F \in C^p(A \otimes \mathfrak{g}, M \otimes \mathfrak{g})$ by

$$F(a_1 \otimes X^1 \wedge \ldots \wedge a_p \otimes X^p) = \hat{f}(a_1 \wedge \ldots \wedge a_p) \otimes X^1 \cdots X^p.$$

A calculation similar to that in the Lemma shows that this map is compatible with the differentials.

## Acknowledgements

I wish to thank Daniel Quillen for suggesting this problem and encouragement.

## References

[1] J.-L. Loday and D. Quillen, Cyclic homology and the Lie algebra of matrices, *Comment. Math. Helv.* **59** (1984) 565–591.

[2] D. Quillen, Algebra cochains and cyclic cohomology, *Publ. Math. I.H.E.S.*, **68** (1989) 139–174.

[3] D. Quillen, Cyclic cohomology and algebra extensions, *K-Theory*, to appear.

MATHEMATICAL INSTITUTE, 24-29 ST. GILES', OXFORD OX1 3LB.

# List of Participants

| | |
|---|---|
| Aitchison, I. | Mathematics, Melbourne, Australia |
| *Alvarez-Gaumé, L. | CERN, Switzerland |
| Aspinwall, P. | Theoretical Physics, Oxford |
| *Atiyah, M.F. | Mathematical Institute, Oxford |
| Banks, D.C. | North Leigh, Oxford |
| Baston, R.J. | Mathematical Institute, Oxford |
| Blencowe, M. | Imperial College, London |
| Brodzki, J. | Mathematical Institute, Oxford |
| Cassa, A. | Mathematics, Trento, Italy |
| Chan, H.M. | Rutherford-Appleton Laboratory, Oxon. |
| †Chisholm, J.S.R. | Mathematical Institute, Kent |
| *Connes, A. | I.H.E.S., France |
| Cornwell, J.F. | Physics, St. Andrews, Scotland |
| †De Vega, H.J. | LPTHE, Paris, France |
| *Donaldson, S.K. | Mathematical Institute, Oxford |
| Esposito, G. | DAMTP, Cambridge |
| Espriu, D.A.R. | Physics, Valencia, Spain |
| †Farwell, R. | St. Mary's College, Twickenham |
| *Goddard, P. | DAMTP, Cambridge |
| de Groot, M. | Theoretical Physics, Oxford |
| Garcia-Prada, O. | Mathematical Institute, Oxford |
| Hannabuss, K. | Mathematical Institute, Oxford |
| †Harvey, W.J. | Mathematics, King's College, London |
| Hawking, S. | DAMTP, Cambridge |
| †Hodges, A. | Mathematical Institute, Oxford |
| Hudson, R.L. | Mathematics, Nottingham |
| Itoh, K. | DAMTP, Cambridge |
| †Kieu, T.D. | Physics, Edinburgh |
| Killingbeck, T. | DAMTP, Cambridge |
| Lancaster, D. | Theoretical Physics, Oxford |
| Lau, Y.-K. | Mathematical Institute, Oxford |
| †Lawrence, R.J. | Mathematical Institute, Oxford |
| Lizzi, F. | Rutherford Appleton Laboratory, Oxon. |
| Loll, R. | Theoretical Physics, Imperial College, London |
| Low, R. | Mathematics, Coventry Polytechnic |

*invited speaker    †contributor

| | |
|---|---|
| Luke, G. | Mathematical Institute, Oxford |
| †MacKenzie, K.C.H. | Mathematical Sciences, Durham |
| Maeda, Y. | Mathematical Institute, Warwick |
| Manojlovic, N. | Theoretical Physics, Imperial College, London |
| †Mansfield, P. | Theoretical Physics, Oxford |
| †Martin, P. | Mathematics, Birmingham |
| *Martinec, E.J. | Enrico Fermi Institute, Chicago |
| Mavromatos, N.E. | Theoretical Physics, Oxford |
| Namazie, M.A. | Rutherford Appleton Laboratory, Oxon. |
| *Narain, K.S. | CERN, Switzerland |
| Nash, C. | St. Patrick's College, Ireland |
| Nellen, L. | Theoretical Physics, Oxford |
| Ortiz, M. | DAMTP, Cambridge |
| Osborne, H. | DAMTP, Cambridge |
| Ozturklevent, H.O. | Mathematics, King's College, London |
| Papadopoulos, G. | Mathematics, King's College, London |
| *Penrose, R. | Mathematical Institute, Oxford |
| *Quillen, D.G. | Mathematical Institute, Oxford |
| †Restuccia, A. | Mathematics, King's College, London, and Caracas, Venezuela |
| Roe, J. | Mathematical Institute, Oxford |
| *Rogers, A. | Mathematics, King's College, London |
| Roland, K. | Niels Bohr, Copenhagen, Denmark |
| †Roscoe, D. | Applied Mathematics, Sheffield |
| Roulstone, I. | Mathematical Institute, Oxford |
| †Sánchez, N. | Observatoire de Paris, France |
| Sarmadi, M.H. | Rutherford Appleton Laboratory, Oxon. |
| Scott, S. | Mathematical Institute, Oxford |
| *Segal, G.B. | Mathematical Institute, Oxford |
| †Singer, M. | Mathematical Institute, Oxford |
| Slodowy, P. | Physics, Liverpool |
| †Taylor, J.G. | Mathematics, King's College, London |
| Tod, K.P. | Mathematical Institute, Oxford |
| *Townsend, P. | DAMTP, Cambridge |
| *Tsou, S.T. | Mathematical Institute, Oxford |
| *Verlinde, E. | Physics, Utrecht, The Netherlands |
| Wang, S. | Mathematical Institute, Oxford |
| Watts, G. | DAMTP, Cambridge |
| Westbury, P.W. | Liverpool |
| Woodhouse, N.M.J. | Mathematical Institute, Oxford |
| Zapowski, A. | DAMTP, Cambridge |